The conservation of lowland dry grassland birds in Europe

Proceedings of an international seminar
held at the University of Reading
20–22 March 1991

Edited by

Paul D Goriup
Secretary, ICBP Steppe and Grassland Birds Group

Leo A Batten
Head of Land Management and
Species Policy Branch, English Nature

and

John A Norton
Nature Conservation Bureau Ltd.

Seminar organised by the Nature Conservation Bureau Ltd. on behalf of the
Nature Conservancy Council and International Council for Bird Preservation.
Report published by the Joint Nature Conservation Committee.

The Joint Nature Conservation Committee and the three country councils carry forward duties
previously undertaken by the Nature Conservancy Council.

Further copies of this report can be obtained from the The Joint Nature Conservation Committee,
Monkstone House, City Road, Peterborough PE1 1JY or The Nature Conservation Bureau Limited,
36 Kingfisher Court, Hambridge Road, Newbury, Berkshire RG14 5SJ.

Copyright © JNCC 1991

ISBN 1 873701 06 3

CONTENTS

Preface	iv
Caroline Jackson MEP	
Recommendations of the Seminar	v
Addendum to Recommendations	ix
List of participants	x
INTRODUCTION, OVERVIEW AND WIDER IMPLICATIONS John Temple Lang	1
THE STATUS OF SEMI-NATURAL GRASSLANDS IN EUROPE Gerard van Dijk	15
THE STATUS OF LOWLAND DRY GRASSLAND BIRDS IN EUROPE Graham M Tucker	37
LAND-USE CHANGES AND THE CONSERVATION OF DRY GRASSLAND BIRDS IN SPAIN: A CASE STUDY OF ALMERÍA PROVINCE Juan Manrique Rodriguez and Eduardo de Juana	49
DRY GRASSLAND BIRDS IN FRANCE: STATUS, DISTRIBUTION AND CONSERVATION MEASURES Patrick Lecomte and Sylvestre Voisin	59
STATUS OF LOWLAND DRY GRASSLANDS AND BIRDS IN ITALY Francesco Petretti	69
STATUS OF LOWLAND DRY GRASSLANDS AND GREAT BUSTARDS IN AUSTRIA Hans Peter Kollar	77
CONSERVATION OF LOWLAND DRY GRASSLAND BIRDS IN HUNGARY Ferenc Márkus	81
THE GREAT BUSTARD IN THE USSR: STATUS AND CONSERVATION Vladimir E Flint and Alexandr L Mishchenko	89
CONSERVATION PROBLEMS OF STEPPIC AVIFAUNA IN TURKEY Y Sancar Baris	93
ACTION FOR DRY GRASSLAND BIRDS IN BRITAIN Richard F Porter, Graham D Elliott and Gwyn Williams	97
IMPLICATIONS OF INTERNATIONAL LAW FOR CONSERVATION OF LOWLAND DRY GRASSLANDS IN EUROPE Richard Buxton	101
IMPLICATIONS OF EC FARMING AND COUNTRYSIDE POLICIES FOR CONSERVATION OF LOWLAND DRY GRASSLANDS David Baldock	111
MANAGEMENT OF SEMI-NATURAL LOWLAND DRY GRASSLANDS John J Hopkins	119
RESTORING AND RE-CREATING SPECIES-RICH LOWLAND DRY GRASSLAND Terry C E Wells	125
RELATED INITIATIVES FOR THE CONSERVATION OF LOWLAND HABITATS Eric M Bignal	133

Preface

This seminar was held at a crucial time for the future of dry grasslands in Europe. The European Community is now being impelled towards a reassessment and perhaps a radical reform of the Common Agricultural Policy. The pressures necessitating this are both budgetary and environmental: the Community budget cannot afford to continue paying expensive subsidies to agriculture, and many Europeans now feel that it would be wrong to continue operating a policy which is so destructive of the rural environment. The topics raised in the seminar were examined in great detail by experts not only from within the European Community but from many countries on its boundaries — some of which may well be members of the Community by the year 2000.

The legislative framework which will ensure a future for Europe's dry grasslands will, in the first instance, have to be that of the European Community. EC directives on nitrates, on habitat protection, on migratory birds — and other proposals still to come — will mean changes in national legislation, and should, in the long term, bring about improvements in conservation status. It will be equally important to ensure that the Common Agricultural Policy is geared far more to financing conservation and environmental protection as a positive means of switching effort from over-production. And once EC laws are in place, we must be prepared to give the European Court of Justice the power to ensure that they are properly implemented and enforced.

As Member of the European Parliament for Wiltshire, I am very conscious that I have an important segment of Europe's surviving dry grasslands in my constituency — but by historical accident rather than design. They have been preserved through the existence of the battle training areas on Salisbury Plain, and would otherwise have disappeared under the plough. We cannot afford to let the process of disappearance by neglect continue. This seminar points the way to much-needed policies of future conservation.

Dr Caroline Jackson MEP

RECOMMENDATIONS OF THE SEMINAR

Contributors: Leo A Batten (English Nature), Paul D Goriup (ICBP Steppe and Grassland Birds Group), Richard Grimmett (ICBP), Stuart Housden (RSPB), Carlos Martin-Novella (Sociedad de Ornitologia Española), Dick Potts (Game Conservancy), John Temple Lang (EC) and Graham M Tucker (ICBP)

The Seminar on the Conservation of Lowland Dry Grassland Birds in Europe was held at the University of Reading from 20-22 March 1991. It was organised by the former Nature Conservancy Council of Great Britain, in association with the International Council for Bird Preservation. The Seminar was attended by 45 participants from 12 countries. While the principal focus of the Seminar was on dry grasslands and their associated birds, the importance of other flora and fauna was appreciated. Moreover, the continuum between dry grasslands, wet grasslands and upland grasslands was explored, especially since many of the causes of habitat loss are common to all grassland types.

DEFINITION

Grasslands considered here are semi-natural and natural grasslands, steppe and other dry grasslands, being virtually treeless plains dominated by non-ericaceous species averaging less than 1 m in height, supporting bird species such as bustards, stone-curlew, partridges and sandgrouse.

ANALYSIS

1. Only remnants of Europe's natural and semi-natural dry grasslands remain. Those that do are of vital importance for at least 42 bird species, of which 27 or more are currently in serious decline in some or all of their range.

2. Dry grasslands are predominantly dependent upon long-established, low intensity farming management. The intensification of farming practices, abandonment or diversion to other land uses, e.g. afforestation, continues to pose a major threat to them.

3. To ensure the conservation of dry grasslands and the flora and fauna which depend upon them it is necessary for European states or regional authorities to integrate conservation objectives into their farming and forestry policies (using methods detailed below).

OBJECTIVES

1. To maintain and enhance remaining dry grassland communities, and seek to re-establish, and where necessary, restore dry grasslands through appropriate management within low-input farming systems.

2. To establish core protected areas to form a biogenetic network under the aegis of competent international organisations. These units must be spread across Europe and be of sufficient size and number to sustain viable populations of grassland flora and fauna.

ACTION BY ALL EUROPEAN STATES OR REGIONAL AUTHORITIES

1. Each state should produce a dry grassland conservation strategy. The strategy should:
- Contain an inventory of all remaining dry grasslands within their territory, this inventory to be regularly reviewed and updated.
- Select priority areas, such as those grassland areas in the 'Important bird areas' (IBA) list (Grimmett and Jones 1989), for urgent conservation action and identify lead agencies to implement such measures. The most important areas for globally threatened bird species are given by Tucker (Table 5 pp. 44-45).
- List the conservation measures required and quantify the funds and resources needed to implement them.
- Require Agriculture and Forestry Departments to support the traditional low-intensity management of dry grasslands.
- Undertake a thorough Environmental Assessment of any significant proposed scheme or development proposal (including publicly funded schemes) which may impact on dry grassland habitats, and take appropriate remedial action to prevent damage.

Land use policies
2. Protect and manage the most biologically important areas through designating national parks, regional parks and other protected areas. Such areas

should be fully protected from inappropriate development, and must include large areas of habitat, as well as small sites for the protection of relic populations, from which the colonisation of surrounding areas could occur.

3. Apply measures such as Environmentally Sensitive (farming) Area (ESA) schemes, to large areas which benefit from existing low-input farming management, and thereby encourage the maintenance of traditional agricultural systems which maintain dry grassland communities. Non-EC States should consider the appropriateness of ESAs and similar schemes for their agriculture. **Priority areas for ESA designation are given in Table 1.**

Re-establishment of dry grassland
4. By the judicious use of long term set-aside, followed by the introduction of sensitive farming, land currently under intensive management could be diverted from its present use so as to restore former areas of dry grassland.

'Extensification'
5. Areas of intensive grassland and crop growing should be considered for extensification measures (low chemical inputs, no irrigation), so as to improve their capacity to support bird species such as bustards.

6. The maintenance of 'low intensity' cereals, or the extensification of cereal production on former dry grassland sites should be considered as a matter of urgency.

7. Dry grassland conservation and management should be encouraged via policies for protecting drinking water supplies (reduction of nitrates), the prevention of erosion and amenity schemes.

THE EUROPEAN COMMUNITY

Protecting known important grasslands
1. Member States should, without delay, classify sufficient grassland areas in the European Community with significant breeding populations of globally threatened grassland bird species (Tucker, this volume, Table 5) as Special Protection Areas (SPA), adequate to ensure their survival and reproduction in their area of distribution. As a minimum, such areas should include those as listed in Grimmett and Jones (1989). Member states should also take measures and avoid deterioration of grassland habitat outside such Special Protection Areas. Management plans for the areas should be drawn up by the appropriate conservation agency and implemented without delay.

Reform of the Common Agricultural Policy
2. CAP reform must recognise the central role that agriculture has in protecting a large proportion of the remaining semi natural and natural dry grasslands in Europe. Financial support to the farming systems which manage dry grasslands must not be eroded by cuts in CAP support, or attacked as "interventionist" or "anti- free trade". Most importantly, CAP must also be made more environmentally benign.

CAP funded schemes must no longer sponsor the absolute destruction of key dry grasslands. Moreover, the market organisations, the mechanism of price policy itself, must be tied to protection of habitats. Support must be directed at those who protect and enhance dry grasslands and withdrawn from those who directly damage them. The recent proposals to reform the CAP (COM 91 100) are novel, by making CAP payments conditional on set aside. This must be extended so that CAP payments become conditional on farmers (a) protecting all wildlife habitats and (b) opting in to one of several schemes, such as extensification, set aside or Environmentally Sensitive Farming.

Environmentally Sensitive Areas
3. For the protection and management of semi-natural dry grasslands, the ESA mechanism as piloted under Article 19 in EC Regulation 797/85, offers substantial opportunities. ESA policy must be developed and expanded so that all semi-natural habitats and farming systems of low intensity qualify for ESA payments.

4. The EC Commission should build on its excellent proposals in COM(90)366, especially those which support existing low intensity farming of great conservation value, by extending the Provisions of Article 19 of Regulation 797/85 to the wider countryside. Careful targeting of ESAs (see Table 1 and Tucker, Table 5) to achieve the maximum conservation benefits is required, with 75% funding

Table 1: Priority areas in the EC for designation as ESAs

Country/Region	IBA Site No
Portugal	
Castro Verde	27
Elvas/Campo Maior Plains	28
Spain	
Villafáfila	38
Madrigal-Peñaranda	43
Plains between Trujillo & Cáceres	112
Brozas-Membrio	115
La Serena	137
France	
Poitou-Charentes Plains	67,71,74,75,76,77,80,82
Italy	
Altipiano di Campeda	114

for those ESAs where special measures to conserve threatened species are needed.

Article 19 should be implemented in countries where significant grassland sites remain, e.g. Portugal, where areas such as Alentejo are high priorities for conservation action.

Other Agricultural Measures

5. Other Agricultural schemes, particularly extensification and set-aside of cereal land, may have an application in the re-establishment of dry grasslands. The maintenance of 'low intensity' cereals, or the extensification of cereal production on former dry grassland sites should be considered as a matter of urgency.

Less Favoured Areas (LFA) Directive

6. The EC (and member states) should review and take the necessary steps to reform the LFA Directive so that it assists, as originally intended, farming which is necessary to protect the countryside and wildlife habitats. Maintaining traditional farming which conserves wildlife habitats should be central to the LFA Directive.

Flora, Fauna and Habitats Directive (Draft)

7. The EC should seek agreement for this Directive without delay, and the importance of semi-natural managed grasslands should be recognised within it and receive urgent action under it. Flora and fauna associated with extensive cereal production, e.g. arable weeds, should be covered by the Directive. Funding to implement the necessary conservation measures must be attached to the Directive.

Structural Funds

8. All current EC-funded projects planned or underway in grassland IBAs or other important dry grassland areas should be the subject of urgent review and Environmental Assessment (EA). This must include examination of forestry projects (e.g. *Eucalyptus* plantations), irrigation schemes etc. Encouragement should be given to projects associated with sustainable use of dry grasslands, e.g. wildlife tourism, 'sweet hay', and other products from low-input farming of such grasslands.

Environmental Assessment

9. Member states which have not already done so should implement Directive 85/337/EEC in respect of agriculture projects listed in Annex II of the Directive, particularly paragraphs 1(b) [*Projects for the use of uncultivated land or semi-natural areas for intensive agricultural purposes*] and 1(d) [*Initial afforestation where this may lead to adverse ecological changes and land reclamation for the purposes of conversion to another type of land use*]. The EC should re-categorise these types of projects in Annex I of the Directive so that they have mandatory EAs. Similarly, regional programmes funded by the EC must be subject to rigorous EA before approval; damage to IBA sites should be avoided.

EUROPE-WIDE INITIATIVES

Research and survey

1. A high priority is to compare and extend the Council of Europe inventory of dry grasslands (Wolkinger and Plank 1981) and produce an authoritative European inventory of dry grasslands. This will need Europe-wide co-ordination but will be reliant on individual states completing national inventories.

A survey programme should be established to determine the European populations of all dry grassland bird species of conservation concern and monitor the effectiveness of conservation measures taken. This is urgently required for the four globally threatened species: *Falco naumanni* (lesser kestrel), *Tetrax tetrax* (little bustard), *Otis tarda* (great bustard) and *Chettusia gregaria* (sociable plover).

Research should be carried out to establish more precisely the effects of agricultural intensification on the flora and fauna of dry grasslands and steppes. In particular, attention should be given to the impact of irrigation, changes in cropping practice, changes in stocking levels and installation of power lines and other infrastructure. Central Iberia and the Anatolian plains of Turkey would be appropriate locations for such projects.

Networking

2. A dry grasslands bureau should be established under the aegis of an existing international organisation, e.g. the International Council for Bird Preservation (ICBP), to maintain a network of specialists to co-ordinate research and survey and to promote the needs of the habitat by way of advice to governments and institutions.

Publicity and Education

3. Publicity and education campaigns directed at decision makers are urgent. These must draw attention to:

- The international importance of dry grasslands and their flora and fauna
- The importance of grasslands in protecting the environment (e.g. combating soil erosion) and for recreation (wildlife tourism, game shooting)
- The catastrophic losses of dry grasslands throughout Europe, and the fact these losses continue
- The need for national inventories of remaining dry grasslands and a concerted programme of action to protect them
- The need for international institutions (e.g. EC) to act quickly to help member states protect their grasslands.

INTERNATIONAL CONVENTIONS

Bonn Convention on migratory species
1. Action by signatories to conserve dry grassland species should be pursued with vigour. As a first step, the agreement to conserve the white stork (*Ciconia ciconia*) should be adopted and implemented, and further agreements applicable to other dry grassland species developed.

Bern Convention
2. The next meeting of the parties to the Convention should be requested to examine the plight of dry grasslands, and encourage signatories to implement national strategies for their protection.

Economic Commission for Europe (ECE)
3. The ECE should be encouraged to promote measures to conserve grassland birds in the context of the Declaration on the conservation of flora and fauna it has issued.

REFERENCES

Grimmett, R.F.A. and Jones, T.A. (1989). *Important bird areas in Europe*. ICBP Technical Publication No. 9. International Council for Bird Preservation, Cambridge.

Wolkinger, F. and Plank, S. (1981). *Dry grasslands of Europe*. Nature and Environment Series, No. 21. European Committee for the Conservation of Nature and Natural Resources, Strasbourg.

ADDENDUM

EVIDENCE PRESENTED BY RSPB TO THE HOUSE OF LORDS COMMITTEE ON THE EUROPEAN COMMUNITIES

BASED ON THE RECOMMENDATIONS OF THE SEMINAR

James Dixon*

The following text was prepared after the Seminar and submitted to the House of Lords European Communities Committee. It is included here because of its relevance to the subject of this Seminar and may be helpful to others who are in similar discussions with their governments.

Future of the CAP

I understand that Sub Committee D is undertaking a wide ranging review of the Common Agricultural Policy and is looking at ways in which the CAP can be reformed, including ways in which environmental policy can be closely integrated with agriculture.

Farming is important for the conservation of many of Europe's most important habitat types. Some of the most important European types of wildlife habitat are semi-natural and natural grasslands, steppe and other dry grasslands. These are characteristically dependent on long established, low intensity farming. They are found across Europe but are best represented in Iberia and central Europe. Only remnants remain, but those that do are essential for the survival of at least 42 bird species, of which 27 or more are currently in decline.

The International Council for Bird Preservation and the (then) Nature Conservancy Council organised a conference on the Conservation of Lowland Dry Grassland Birds in Europe at Reading University from 20-22 March 1991. The seminar was attended by 45 participants from 12 countries and focused on the conservation of birds and other flora and fauna associated with a wide range of grassland types. The NCC, in conjunction with the ICBP, have published a seminar compendium, a copy of which I enclose. The seminar also requested a panel to prepare a set of recommendations. The recommendations will be sent to national authorities responsible for agriculture and nature conservation and to the European Commission.

The recommendations panel felt strongly that its views relating to agricultural reforms should be taken into account by the Sub-Committee as they are now investigating this topic. The panel made seven recommendations to Member States and Regional Authorities of which five were specifically agricultural, and nine recommendations to the European Community of which six were specifically agricultural. Therefore, I have listed below the conservation objectives for dry grassland and the main recommendations relating to **agricultural policy**.

Conservation Objectives

i To maintain and enhance remaining dry grassland communities, and seek to re-establish, and where necessary, restore dry grasslands through appropriate management with low-input farming systems.

ii To establish core protected areas to form a biogenetic network under the aegis of competent international organisations. These units must be spread across Europe and be of sufficient size and number to sustain viable populations of grassland flora and fauna.

Action required by all European States or regional authorities:

1. Produce a dry grassland conservation strategy which should:
   - contain an inventory of all remaining grasslands,
   - select priority areas for conservation action and identify lead agencies to implement such measures,
   - list the conservation measures and quantify the resources needed,
   - **require agriculture and forestry departments to support the management of traditional low intensity dry grasslands,**
   - undertake an environmental assessment of proposed works, including agricultural, which may impact on dry grasslands.

2. Apply measures such as Environmentally Sensitive Area (ESA) schemes, to large areas which benefit from existing low-input farming management, and thereby encourage the maintenance of traditional agricultural systems which maintain dry grassland communities. Non EC States should consider the appropriateness of ESAs.

3. By the judicious use of long term set-aside, followed by the introduction of sensitive farming, land currently under intensive management could be diverted from its present use so as to restore former areas of dry grassland.

4. Areas of intensive grassland and crop growing should be considered for extensification measures (low chemical inputs, no irrigation), so as to improve their capacity to support bird species such as bustards. The maintenance of "low intensity" cereals, or the extensification of cereal production on former dry grassland sites should be considered as a matter of urgency.

5. Dry grassland conservation and management should be encouraged via policies for protecting drinking water supplies (the reduction of nitrates), the prevention of erosion and amenity schemes.

Action required by the European Community

1. CAP reform must recognise the central role that agriculture has in protecting a large proportion of the remaining semi natural and natural dry grasslands in Europe. Financial support to the farming systems which manage dry grasslands must not be eroded by cuts in CAP support, or attacked as "interventionist" or "anti-free trade". Most importantly, CAP must also be made more environmentally benign.

CAP funded schemes must no longer sponsor the absolute destruction of key dry grasslands. Moreover, the market organisations, the mechanism of price policy itself, must be tied to protection of habitats. Support must be directed at those who protect and enhance dry grasslands and withdrawn from those who directly damage them. The recent proposals to reform the CAP (COM 91 100) breaks new ground by making CAP payments conditional on set aside. This must be extended so that CAP payments become conditional on farmers (a) protecting all wildlife habitats and (b) opting in to one of several schemes, such as extensification, set aside or Environmentally Sensitive Farming.

2. For the protection and management of semi-natural dry grasslands, the ESA mechanism as piloted under Article 19 (EC Council Regulation 797/85), offers substantial opportunities. ESA policy must be developed and expanded so that all semi-natural habitats and farming systems of low intensity qualify for ESA payments.

3. The EC Commission should build on its excellent proposals in COM (90)366, especially those which support existing low intensity farming of great conservation value, by extending the provisions of Article 19 to the wider countryside. Careful targeting of ESAs to achieve maximum conservation benefits is required, with 75% funding for those ESAs where special measures to conserve threatened species are needed. The principle of subsidiarity should apply; Member States should be **required** to implement schemes in order to achieve the conservation of species or habitats recognised (under EC Directives) as under threat. It would then be **optional** for Member States to implement more widely available schemes for more local concerns.

4. Other agricultural schemes, particularly extensification and set aside of cereal land should be used to benefit nature conservation to the maximum. The provision for premium payments to "top-up" set aside and the extension of set aside to agreements lasting up to 20 years (as proposed in COM (90)366) may have an application in the re-establishment of dry grasslands.

5. The EC should review and take the necessary steps to reform the LFA Directive so that it assists, as originally intended, farming which is necessary to protect the countryside and wildlife.

6. All current EC funded projects planned or underway in grassland important bird areas (IBAs) should be the subject or urgent reviews and Environmental Assessment. This must include forestry projects and agricultural projects such as irrigation and drainage, funded by CAP budgets.

*Agricultural Policy Officer, Royal Society for the Protection of Birds, The Lodge, Sandy, Beds SG19 2DL, UK

LIST OF PARTICIPANTS

Mr David Baldock
IEEP
3 Endsleigh Street
London WC1H 0DD
UK

Dr Leo A Batten
English Nature
Northminster House
Peterborough PE1 1UA
UK

Dr Eric M Bignal
JNCC
Kindrochaid
Bruichladdich
Isle of Islay
Argyll PA44 7PP
UK

Mr G Boobyer
JNCC
Monkstone House
City Road
Peterborough PE1 1JY
UK

Ms Margaret Brown
Berks, Bucks & Oxon Naturalists' Trust
3 Church Cowley Road
Rose Hill
Oxford OX4 2JR
UK

Ms Nicola Crockford
JNCC
Monkstone House
City Road
Peterborough PE1 1JY
UK

Dr Eduardo de Juana
Sociedad de Ornitologia Española
Facultad de Biologia PL9
Ciudad Universitaria
28040 Madrid
SPAIN

ir Gerard van Dijk
Roerdompstraat 31
3815 BT Amersfoort
NETHERLANDS

Mr Jim Dixon
RSPB
The Lodge
Sandy
Beds SG19 2DL
UK

Prof V E Flint
All Union Research Institute of Nature Conservation
Znamenskoye-Sadki
113628 PO Vilar
Moscow
USSR

Dr Colin Galbraith
JNCC
Monkstone House
City Road
Peterborough PE1 1JYUK

Dr Martin George
Marsh House
Tinkers Lane
Strumpshaw
Norwich NR13 4HT
UK

Mr David Glaves
MAFF
Stap Lake Mount
Starcross
Exeter EX6 8PE
UK

Mr Paul D Goriup
Nature Conservation Bureau Ltd.
36 Kingfisher Court
Hambridge Road
Newbury RG14 5SJ
UK

Mr Richard Grimmett
ICBP
32 Cambridge Road
Girton
Cambridge CB3 0PJ
UK

Dr John J Hopkins
JNCC
Monkstone House
City Road
Peterborough PE1 1JYUK

Mr Stuart Housden
RSPB
The Lodge
Sandy
Beds SG19 2DL
UK

Mrs R Howard
Child-Beale Wildlife Trust
Church Farm
Lower Basildon
Reading RG8 9NH
UK

Mr E T Idle
English Nature
Northminster House
Peterborough PE1 1UA
UK

Dr Caroline Jackson
74 Carlisle Place
London SW1P 1HZ
UK

Mr Miles King
Berks, Bucks & Oxon Naturalists' Trust
3 Church Cowley Road
Rose Hill
Oxford OX4 2JR
UK

LIST OF PARTICIPANTS

Dr Hans Peter Kollar
Institut Fur Angewandte Oko-Ethologie
Kirchengasse 34
2285 Leopoldsdorf
AUSTRIA

Patrick Lecomte
GEPANA
26 Rue D'Estienne D'Orves
92120 Montrouge
FRANCE

Dr Heinz Litzbarski
Dorfstrasse 9A
1831 Nennhausen
Brandenburg
GERMANY

Mr Ferenc Márkus
Koltu U.21
H-1121 Budapest
HUNGARY

Snr Carlos Martin-Novella
Sociedad de Ornitologia Española
Facultad De Biologica PL9
Ciudad Universitaria
28040 Madrid
SPAIN

Ms Claire McHardy
Berks, Bucks & Oxon Naturalists' Trust
3 Church Cowley Road
Rose Hill
Oxford OX4 2JR
UK

Dr A L Mishchenko
All Union Research Institute of Nature Conservation
Znamenskoye-Sadki
113628 PO Vilar
Moscow
USSR

Mr John A Norton
Nature Conservation Bureau Ltd.
36 Kingfisher Court
Hambridge Road
Newbury RG14 5SJ
UK

Dr Coleman Ó'Críodáin
National Parks and Wildlife Service
2 Sidmonton Place
Bray
County Wicklow
IRELAND

Snr Francesco Petretti
WWF Italia
via Salaria 290
00199 Roma
ITALY

Dr Mike Pienkowski
JNCC
Monkstone House
City Road
Peterborough PE1 1JY
UK

Ms Marcia Pinto
SNPRCN
R Filipe Folque 46 3'
1000 Lisboa
PORTUGAL

Mr Richard F Porter
RSPB
The Lodge
Sandy
Beds SG19 2DL
UK

Dr G R Potts
The Game Conservancy
Fordingbridge
Hants SP6 1EF
UK

Mr Mick Rebane
RSPB
The Lodge
Sandy
Beds SG19 2DL
UK

Dr John Temple Lang
Ave P Hymans 113, BTE19
B1200 Brussels
BELGIUM

Ms Liz Thomas
Berks, Bucks & Oxon Naturalists' Trust
3 Church Cowley Road
Rose Hill
Oxford OX3 2JR
UK

Dr Graham M Tucker
ICBP
32 Cambridge Road
Girton
Cambridge CB3 0PJ
UK

Dr Hilary Tye
WCMC
219c Huntingdon Road
Cambridge CB3 0DL
UK

Sylvestre Voisin
GEPANA
26 rue D'Estienne D'Orves
92120 Montrouge
FRANCE

Dr Terry C E Wells
Monks Wood Experimental Station
Abbots Ripton
Huntingdon
Cambs PE17 2LS
UK

Mr Gwyn Williams
RSPB
The Lodge
Sandy
Beds SG19 2DL
UK

INTRODUCTION, OVERVIEW AND WIDER IMPLICATIONS

John Temple Lang

Director, Directorate General for Competition, Commission of the European Communities*

Former Chairman, ICBP-Ireland and ICBP European Continental Section

INTRODUCTION

Everyone is familiar with the former extent of the North American prairies, but large areas of grassland are not thought of as part of Europe's natural vegetation. Indeed, since the immediate post glacial period the largest grassland areas in the Western Palearctic have been the steppes in the USSR, the *puszta* in Hungary, and the Anatolian plateau, and there is some doubt about how far these were natural, or would change to scrub or woodland if left unmanaged today. Yet the large number of invertebrate species associated with European grasslands, and occurrence of fossil bones of great bustard *Otis tarda* at many sites in Europe, suggest that natural grassland was much more widespread than it is today (Van Dijk, this volume). Because so little grassland in Europe is considered natural, and so much of what remains is obviously altered by man or domestic stock, the scientific and conservation importance of European grasslands has been underestimated, certainly by ornithologists (Goriup and Batten 1990). It was one of the aims of this seminar to correct this. Also, the ecology of agricultural areas is not as well known as its importance requires, and the seminar set out to collect the information available on grassland birds, as relevant to conservation.

The title of the seminar was "the conservation of lowland dry grassland birds in Europe". Several comments are needed about the scope of this subject. First, it was agreed in discussion that "lowland" is not appropriate to the grasslands of the Spanish plateaux at 1,600 m and 1,900 m which have similar ecological characteristics, and similar bird species, to lower-lying grasslands further east (Rodríguez and de Juana). The same can be said of many secondary grasslands in mountains and in the alpine zone where we find important areas of grasslands which have become rare elsewhere (e.g. in the Jura). Second, "dry" grasslands are at one end of a phytosociological spectrum: they are not a distinct category, and botanists at the seminar felt that the conclusions should apply to other grassland types which are as scientifically interesting and as seriously threatened. The ornithologists present however felt that the scope of the papers and the discussion, and the aims of the seminar, had been limited to the drier grassland habitats. They would not have been able, without much further discussion, to extend the conclusions of the seminar to all other kinds of grassland areas, unless they had merely repeated conclusions reached at three earlier meetings on the chough *Pyrrhocorax pyrrhocorax*, a Wader Study Group workshop, and a European Forum on Birds and Pastoralism (Bignal). They pointed out that while botanists can undoubtedly list some very important European grassland areas such as the Jura mountain meadows, and the Loire and Bug riverine grasslands, there does not seem to be an authoritative Europe-wide list of botanically important areas, or even of important grasslands, corresponding to the ICBP list of important areas for birds in Europe.

A third comment on the scope of the seminar is that the bird species which use European dry grasslands are not a clearly defined group of species or a single bird community (the word "guild" was never mentioned during the seminar). Much of the discussion naturally concentrated on species which are confined to or dependent on dry grassland. But of these, even the four globally threatened species, great bustard, little bustard *Tetrax tetrax*, lesser kestrel *Falco naumanni* and sociable plover *Chettusia gregaria* use arable land in some circumstances. Other species such as sandgrouse *Pterocles* spp. and some larks Alaudidae are more naturally described as semidesert species rather than grassland species, and do not inhabit the same kinds of ecosystems, as for example,

---

*Address for correspondence: Ave P Hymans 113, bte 19, B-1200 Brussels, Belgium

pipits *Anthus*. Some other species are not confined to or dependent on grassland, but need open areas, and use grasslands when they are available. These species include insectivorous species such as swifts *Apus* spp. and hirundines, bee-eater *Merops apiaster*, roller *Coracias garrulus*, several species of pipit (Petretti) and carnivorous species such as vultures and other diurnal raptors. Bird species are more tolerant than plants of ecological variations, and their characteristic habitats are less precisely defined.

Another point, important for conservation as well as for scientific reasons, is that some European birds use dry grasslands as staging points on migration or for wintering, even though they do not nest in dry grassland: these include cranes *Grus grus* and white stork *Ciconia ciconia*.

References were made in discussion to "flagship species" regarded as indicators of the high quality of the ecosystems in which they occur. The concept needs clarification. First, the phrase should be used as an ecological concept, not merely a public relations one (there is a tendency to use it in connection with striking species such as bustards, rather than larks for example, flora or invertebrates). Second, we probably do not know enough about the requirements of grassland bird species to say whether the presence of a species proves anything precise about the habitat. Great bustards for example use arable land which is not a natural ecosystem and which may be suboptimal for the species itself. Presence might mean little more than there is a (stable) population of the bird in question in the area. Third, when speaking of species adapted to anthropogenic ecosystems, it is not clear what is meant by "high quality" of the ecosystems. The main requirements of chough are apparently short, grazed coastal or montane grass and abundant droppings of domestic livestock: the fact that these requirements are met and that the birds are present does not necessarily prove anything else about the floristic or faunistic richness of the area. If we knew enough about their requirements, the presence of certain species of flora or invertebrates, though less striking, might be a better indicator of ecological state of the area than the presence of any bird species. Birds after all can and apparently do compensate for the inadequacies of one ecosystem by moving to others, while plants cannot do so.

As far as conservation of bird species is concerned, the situation described by the seminar papers is unexpectedly bad. Paper after paper described declining bird populations in habitats decreasing in area and declining in quality. The common unspoken assumption of northern European ornithologists that bird populations remain at satisfactory levels in regions of low-intensity agriculture were shown to be wrong, in spite of lack of precision in older count data.

The reasons why the seriousness of the situation have been underestimated seem to be:

(1) Most dry grassland bird species are (and for ecological reasons must be) dispersed over large area. They are therefore difficult to count, and few ornithologists until recently have counted them. In the most important areas, such as Turkey, Spain and the USSR, there are not enough ornithologists to count the large areas involved thoroughly from the ground, and aerial counts are unsatisfactory.

(2) In some areas at least the declines are accelerating. Indeed, in some respects even the seminar papers seemed on occasion to underestimate the seriousness of the present situation because they were based on some out of date information (e.g. concerning Ireland).

(3) A third reason, more important for flora than for birds, is that artificial fertilisers are among the most important elements causing change in grasslands, and their effects easily go unnoticed.

(4) Some species have moved into arable land, and it has not been understood until recently that this might lead to decreased reproductive success.

(5) Conservationists, scientists and the general public alike tend to regard man-made habitats such as grassland as less interesting than natural ecosystems.

(6) Dry grassland species have not been considered important priorities for bird conservation (Petretti), even in countries where the conservation movement is strong (Porter, Elliot and Williams).

The main threats to dry grasslands and their birds are similar throughout Europe (and, as Van Dijk and others pointed out, apply also to other kinds of grasslands). They are more intensive agriculture, conversion to tillage, fertilization, use of herbicides and insecticides, irrigation, over-grazing (notably in Turkey), abandonment, and afforestation or conversion to olive and almond groves, and vineyards. Other threats include undergrazing (leading to ecological succession to scrub), fire, soil erosion, disturbance (due to e.g. military activities), urbanisation and in industrialisation, lowering of the water table by water extraction, fragmentation of fields and (in Hungary) creation of fishponds and ricefields (Márkus). As Baris was unable to attend the seminar, the effects of the huge Southeast Anatolia Project (GAP) dam were not adequately discussed (see Baris, this volume).

The papers prepared for the seminar fall into four groups, and it is convenient to introduce them here in the following order. There are two general papers on grasslands and grassland birds in Europe; country reports; three papers on management and restoration of dry grassland and related initiatives, and two papers on implications of present international law and future official policies on agriculture for conservation.

THE PAPERS

Grasslands and grassland birds in Europe

From a botanical perspective, Van Dijk calls attention to the loss of genetic resources of fauna and flora which has already occurred as a result of the decline in area and in quality of all kinds of grasslands in Europe in this century. The decline in grassland bird populations, though serious, is less than the "massive" decline of semi-natural vegetation and related invertebrates. He gives estimates of the small and fragmented areas of grassland remaining in a number of European countries, and rightly stresses that small sites, many of them linear, are not an adequate substitute either for nature reserves or for appropriated conservation policies in multiple-use or farming areas outside reserves. Nature reserves are essential for some species which cannot survive any modern agriculture, and reserves can be a genetic bank for future colonisation of surrounding areas when restoration occurs. In Europe as a whole, France, Ireland, Poland and Yugoslavia hold the most important grasslands, but an overall assessment is difficult due to a lack of information for some potentially important countries including the USSR, Spain, and Turkey.

Tucker identifies 27 out of 42 species of lowland dry grassland birds in Europe which he regards as being "of conservation concern"; most of these are declining, and the four species mentioned earlier are threatened with extinction world-wide. The importance of populations of dry grassland birds in Spain, Turkey and the USSR is due to both the extent of grassland and the relatively low intensity agriculture there. A series of important tables show population trends and conservation priorities for the 42 species in question, and the protected or unprotected status of grassland areas listed by ICBP (Grimmett and Jones 1989) with nationally or internationally significant breeding populations of the four globally threatened species.

In his paper, Tucker also draws a very important distinction between areas which are actual or potential nature reserves and areas which, because of their size, or for other reasons, cannot be made into nature reserves and therefore can only be conserved by multiple-use policies taking account of the needs of wildlife. Probably most of the areas in Spain listed by Grimmett and Jones (1989) are both important enough and (relatively speaking) "small" enough to justify nature reserve status.

It had been assumed that the USSR supported large areas of grasslands, particularly dry grassland steppes, which were of enormous importance in Europe as a whole. However, Van Dijk (pers. comm.) is sceptical about how much dry grassland actually remains. His data was derived from IUCN (1990) where the grassland types were not categorized (into wet or dry). At the seminar, Flint reported that steppe biotopes have become rare, and relatively small in size. Although Grimmett and Jones list some areas in the ten western republics of the USSR, this list is incomplete and unsatisfactory as far as lowland dry grasslands are concerned. The region northeast of the Black Sea is the largest area of steppe west of the Ural mountains. The most interesting areas are apparently in the east of the Ukraine, in the south of the Russian republic and in Kazakhstan. Although a relatively small proportion of this huge region is still virgin steppe, there are several large and immensely important nature reserves, including:

- Askania-Nova, near the mouth of the river Dniepr on the Black Sea coast (area 111 km^2)
- Tsentralno-Chernozomny (48 km^2)
- Lugan, near Voroshilovgrad, Ukraine (15.8 km^2)
- Ukrainsky (16 km^2)
  (Knystautas 1987, Curry-Lindahl 1964).

Of these, only the first and the fourth are listed in Grimmett and Jones (1989). IUCN (1990) give a figure for the total area of protected grasslands in European USSR as 1.991 million ha (19,910 km^2).

Country reports

The country reports all tell similar stories, in fascinating detail. In Almería, Rodríguez and de Juana state that land use changes have led to a decline in most grassland bird species. Several of the species involved are already on the Spanish Red List. We lack information about the grazing pressure needed to maintain rich and diverse steppe bird communities.

In Italy, Petretti reports that although true Mediterranean steppes have been drastically reduced, there are still areas which are important for various species, in particular little bustard, which shows a strong preference for permanent grazed pastures. Conservation of steppe depends on extensive sheep farming.

Kollar says that in Austria the great bustard was widespread in the open areas in the Pannonic east of Austria, but these are now too small to be suitable and the species nests only on agricultural land, with reduced reproductive success. In spite of very careful management for the bustards, only 25 to 28 individuals remain.

Even the extensive Hungarian *puszta* grasslands have been shrinking. They were maintained by grazing, Márkus says, and now large areas are put to other uses, and are fragmented and of poor quality ecologically. Optimal grazing methods, numbers, species and breeds of livestock need to be assessed: "reserve-orientated nature conservation" is not enough.

Turkish steppes are inhabited by 96 breeding bird species from 65 genera, Baris writes, and are used for hunting and foraging, and in winter on migration, by many other species. Overgrazing is the most important single problem: grazing pressure has doubled between 1950 and 1985. Research on populations and distribution, and on the likely effects of the huge dam in southeast Anatolia, are urgently needed.

Lecomte and Voisin reported that several dry grassland birds had become extinct in France, and that 18 of the French grassland bird species are on the French Red List. Most of the threats are from changing farm practices.

Six of the grassland species in Britain discussed by Porter, Elliott and Williams are threatened, mainly by agricultural intensification and related factors. However, grazing is needed to maintain conditions suitable for stone curlew *Burhinus oedicnemus* and woodlark *Lullula arborea*, and maintenance of low intensity agricultural is desirable for other species.

Management, restoration and conservation of dry grasslands

The five remaining papers are more concerned with conservation-orientated measures of various kinds. In the first of three botanical papers on management and conservation, Hopkins says that much of what has been written about management of grassland has concerned areas where nature conservation is the only, or primary aim. He analyses grassland management in agricultural areas, but stresses that in Britain at least grassland plant communities vary considerably in the management they need. Vegetation structure is important, and different species need different structures. Hopkins and Wells both stress that addition of nutrients reduces floristic diversity, and if substantial enrichment occurs it may be difficult to restore the area to its former state. The different effects of grazing, cutting and burning are described. The types of farming compatible with nature conservation are not necessarily the most profitable, and they are therefore specially vulnerable to economic and policy changes. Policies need to consider semi-natural grassland as a management system as well as an ecosystem, Hopkins concludes, and to protect the features of farming systems which are compatible with nature conservation. The chances of creating new grasslands successfully are limited until existing management problems have been solved, so preventing habitat destruction is the most immediate priority.

Wells describes techniques for restoring and recreating grasslands which are species-rich in flora, while admitting that much more research is needed to adapt agricultural methods of management to nature conservation purposes. Soil fertility is an important factor. Some of his suggestions for further research are mentioned below.

International law and policies in Europe

Under the headings of ecological relationships, regional variations in farming systems, and official policies, Bignal summarises proposals made at three recent meetings concerned with various aspects of conservation of grassland birds outside nature reserves. Permanent pasture grassland in low-intensity farming systems have richer invertebrate faunas and more complex vegetational structures than intensively managed land. Long term monitoring of breeding birds is needed: some birds may be attracted to breed early in the season to nest in areas where subsequent chick survival is low. Conservation proposals must seek to maintain low density farming systems if management is to have long term benefits. Agricultural policy makers, he says, still fail to accept the environmental and social benefits of traditional agriculture.

What can conservationists do in the light of all this? The answers can be summarised under three headings; they can:

- use existing international and national legal measures to conserve grassland regions, and develop new legal measures if necessary (Buxton)
- try to influence future agricultural policies, in particular those of the European Community, which are now under radical reconsideration, in ways consistent with nature conservation (Baldock)
- promote research (several authors)

I first comment on the papers of Buxton and Baldock. The implications of these two papers are not limited to grasslands or even to birds. Indeed, since most species of European fauna and flora must be conserved at least partly outside nature reserves, these papers raise issues relevant to the conservation interest of almost every wild species in Europe. I have therefore added some wider considerations, and I then turn to research.

Conservation by existing legal means: comments on Richard Buxton's paper

This paper summarises the Bern and Bonn Conventions, and two European Community (EC) measures, the Birds Directive and the Environment Assessment Directive. To assess the relative importance of these measures, several points should be kept in mind:

1. Contrary to what Buxton says, the obligations in the Berne Convention in relation to habitat conservation are **not** "clear". It merely states:

"Each Contracting Party shall take *appropriate and necessary* legislative and administrative measures to ensure the *conservation* of the *habitats* of the wild flora and fauna species, *especially* those specified in the Appendices I and II, and the *conservation* of *endangered* natural habitats.

The Contracting Parties in their planning and development policies shall *have regard* to the conservation *requirements* of the areas protected under the preceding paragraph, so as to avoid *or minimise as far as possible* any deterioration of such areas.

The Contracting Parties undertake to give *special attention* to the protection of areas that are of importance for the migratory species specified in Appendices II and III and which are *appropriately* situated in relation to migration routes, as wintering, staging, feeding, breeding or moulting areas.

The Contracting Parties undertake to coordinate as *appropriate* their efforts for the protection of the natural habitats referred to in this article when these are situated in frontier areas."

All the words I have put in italics are imprecise both legally and scientifically. Progress under the Berne Convention has been disappointing (Batten 1987). Apart from the vagueness of the words italicised above, this may be partly because the lists of species in the appendces were not well-considered. Too many species are listed as deserving "special" protection, and the effect of this is to reduce the value of the obligation to conserve habitat. Other weaknesses of the Convention, in comparison with the EC measures, are that there is no Court to enforce it, no independent secretariat to develop and supplement it, and no funds to help with the cost of habitat conservation. The mechanisms so far used to clarify the obligations under the Convention (resolutions defining or making more precise some important concepts) have not cured these weaknesses. Buxton also mentions the lack of reporting by States about their compliance with the Convention. Tucker rightly says that the Convention has had little impact on dry grassland birds.

2. The EC Birds Directive, and the legally binding Resolution which accompanied it are enforceable, when necessary, in the Court of Luxembourg. Some funds are available (under the Action of the Community on the Environment programme) for action to implement the Directive, including habitat conservation and improvement. The Directive imposes more precise obligations than the Berne Convention, and the Resolution requires Member States to list the areas which they have classified in accordance with the habitat protection provisions of the Directive (Article 4), the areas which they have or intend to have designated as wetlands of international importance, and other areas protected under national law. In addition, the EC Commission is obliged to coordinate national measures to see that the network of reserves fulfils the objectives of the Directive "and can be integrated into a larger network, should the need arise". The Commission was also to develop criteria for selecting and administering special protection areas. Unlike other international conservation measures, the Directive does not leave it to the discretion of States whether to conserve any particular area. If an area fulfils the criteria referred to in the Directive and the Resolution, it should be conserved. Also, the EC Commission has obtained from the International Waterfowl and Wetlands Research Bureau (IWRB) and ICBP an authoritative list of the most important areas for birds in the twelve EC Member States, which can be used to assess whether the areas protected or proposed for protection by national authorities are sufficient to ensure that the aims of the Directive are fulfilled.

A series of cases have been brought by the Commission before the Court to ensure compliance with the Directive. The most important case for habitat conservation concerned the Scottish blanket bog area, Duich Moss. Under pressure from the court proceedings, steps were taken ultimately to protect this area without a judgment of the Court being necessary, but the case shows that EC Directives are far more effective than Conventions (Temple Lang 1982, 1990).

The areas listed by IWRB and ICBP for the EC Commission are included in Grimmett and Jones (1989) and of course include a number of lowland dry grasslands. The EC Commission is actively concerned with the application of the habitat provisions (as well as the other provisions) of the Directive.

3. Buxton says that the Directive on environmental impact assessments is "the most effective legal instrument". Experience will show if this view is correct: the national measures needed to implement it have mostly been in force only since June 1988. One should recognise however that the Directive only requires an assessment to be made before projects with substantial effects on the environment are carried out. It does not impose any substantive obligations to modify or prohibit those which are found to have undesirable effects. Also, in practise, much of the information on which assessments are based in practice from the developers, and its objectivity can hardly be relied on. Much of the effectiveness in practice of the Directive will depend on the ways in which the national implementing legislation is interpreted and applied. Under European Community law all national measures giving effect to a Directive *must* be interpreted so as to give effect to the Directive as far as possible. The Directive obliges the responsible authorities to take the information resulting from the Directive into consideration: it does not, and in the nature of things could not, oblige them to act correctly or wisely on the basis of that information. However, obligations under e.g. the Birds Directive to conserve particular areas of scientific importance are harder to ignore, due to the Environmental Impact Assessment directive.

The Directive has the effect of ensuring that the implications of the proposed development are spelled out in detail, in the light of any international obligations applicable to the area in question. This obliges the authors of the assessment to say whether and to what extent the development is inconsistent with the applicable obligations, and whether and if so how any conflict between them can be minimised or avoided. The Assessment Directive is, therefore, a useful tool, provided that the area affected has already been identified as important and provided that enough is known about its ecology for the effects of the development to be foreseen with reasonable confidence.

4. The final terms of the EC Habitats Directive are not yet known, but it seems clear that it will be linked to funds for its implementation, and it will of course cover ecosystems and organisms other than birds. It will also contain more precise obligations in relation to specific sites or types of sites than the Birds Directive, although these obligations are precisely those which have given rise to the greatest difficulty.

As far as can be derived from the appendices (Buxton, this volume), only abandoned grasslands are included, whereas "extensive" (low input) use is essential to maintain their quality (except for those abandoned grasslands that are managed by nature conservation bodies).

5. Although Directives under EC law require to be implemented by whatever national measures are appropriate in the light of the previous national laws and procedures, in any litigation in national course between an EC Member State authority and any private person, the State is bound by the terms of any relevant Directives. It cannot therefore rely on or take advantage of the fact that it has not implemented the Directive adequately or at all: it must, in effect, be treated by the national court as if it had done so, when the obligations imposed are clear and leave the State no choice as to how to carry them out.

6. It should be stressed that the national laws of many European countries give various possibilities, which have not yet been fully explored or utilised, to private persons to bring legal proceedings to compel public authorities to fulfil their obligations to conserve the environment, whether the obligations result from national laws or from international or European Community measures. Experience from the United States has shown that proceedings of these kinds can be effective weapons in the hands of determined conservation bodies represented by imaginative lawyers.

7. Buxton says that "pressure" from private interests is needed to ensure fulfilment of States' obligations under international measures. The contribution which private bodies can make is considerable, if (but only if) they are sufficiently professional and industrious. A well-organised conservation body should be capable of preparing a well-documented case, in effect doing the work of the enforcing authority. This not only saves time and makes less demands on the authority's (usually scarce) manpower. It also ensures that the authority cannot dismiss a complaint as unproven or insufficiently important to need investigation. It is a common assumption of conservation bodies in Europe that lobbying rather than legal action is the most effective means available to them: this assumption is untrue, and it weakens conservation efforts, and is less than professional. What conservation bodies *should* be doing is discussed below.

8. The Bern Convention is open to accession by any State which may be invited by the Council of Europe to accede. It is not limited to present or even to possible future members of the Council of Europe. However, the lists of species which form the Appendices to the Convention were drawn up with Council of Europe States in mind, and indeed do not seem well-conceived even in relation to e.g. Turkey, a long standing member of the Council of Europe.

This point is relevant to the question whether it would be useful, in order to ensure conservation of lowland dry grasslands in States not now members of the Council of Europe, that they should be encouraged to become parties to the Bern Convention.

9. The Bonn Convention on the Conservation of Migratory Species of Wild Animals, which covers birds, is in principle a worldwide Convention: any State anywhere can become a party to it. This Convention obliges the parties to it inter alia to try to conserve and if necessary to restore habitats which are important to the survival of a relatively short list of endangered species, if the State in question is within the range of the endangered species. The Convention is essentially a rather elaborate basis for a series of supplementary international agreements between States whose territories form the range of a species or a group of species which are in need of conservation. Guidelines for the terms of these supplementary agreements are set out in the Convention.

It has to be said that progress in drawing up these supplementary treaties, without which the Bonn Convention is of limited importance, has been disappointedly slow. No supplementary agreement has yet been concluded, although an agreement for the conservation of white stork is under discussion, as well as a much more ambitious agreement for the conservation of West Palaearctic geese and ducks. It has been suggested (Goriup and Batten 1990) that the Agreement on the white stork "could serve as a basis for a wider Agreement" for lowland grassland species.

The Bonn Convention is primarily intended to be the basis of active cooperation between States whose territories comprise the ranges of migratory species. It is less appropriate as a basis for cooperation e.g. by exchange of information or techniques, between States each of which holds a sedentary population of an endangered species, or which hold separate populations which migrate little. The Bonn Convention provisions and procedures are over-elaborate when a single population does not need to be managed by several States together making up the range of the species.

10. Environmental Impact Assessment legislation, whatever its precise strengths and weaknesses in a given country, is not limited to birds or to grasslands: it offers opportunities for making conservation arguments across the whole range of physical planning situations. Legal knowledge, techniques, and experience developed in Impact Assessment procedures can be used for the conservation of any ecosystem or species. The cost of getting this knowledge and experience will be repaid many times. Although national Impact Assessment regimes will have special features, they will mostly be similar to the European Community model, because even States not yet members of the EC are planning to join as soon as their economic development permits. This means that conservationists with experience of Impact Assessment controversies can advise and help conservationists elsewhere with less experience of this kind. Conservation organisations should try to arrange this kind of cooperation.

What can international conservation measures do?
Habitat conservation is a necessary condition, though not always a sufficient condition, for conservation of most species. A State which wants to conserve the habitat of a given species, or an area of scientific interest, on its territory, does not need to make a treaty in order to do so., So it is important to see clearly what functions international conservation measures can perform. They can:
- ensure that conservation measures in one State are not made futile by the inaction or actions of another state.
- facilitate international coordination of conservation measures along migration routes
- facilitate exchange of information and conservation know-how.
- provide a way of putting pressure on States which have taken no conservation measures, or inadequate measures.
- provide authorities with an excuse for taking conservation measures which are necessary but unpopular.

Therefore, when an international conservation measure is proposed, or when it is suggested that an existing international measure should be used for some new purpose, it is important to be clear precisely what is desired, and how it can best be achieved. This means, among other things, that it is not necessarily useful to lobby governments to conserve particular areas unless one knows that conservation measures are really needed to maintain or restore the habitat there.

Sometimes a State may become party to an international conservation treaty for reasons not directly related to conservation, e.g. to prove that it is a modern, environmentally-minded country, or to exercise its treaty making powers in a new area. Provided that the State concerned can be persuaded to carry out its treaty obligations, and is not concerned only with public relations, conservationists can take advantage of such a situation. It is wise to be clear about what is really going on.

Another situation arises when a State is asked to become a party to a conservation treaty in order to persuade it to give weight to conservation considerations equal to the weight it already gives to environmentally damaging policies. Although this is an important type of situation, it is really only a special case of using a treaty to get a State to take measures which it would not otherwise take.

Another specific situation which is worth mentioning is that a conservation lobby may persuade its own State to become a party to a conservation treaty so that the lobby can obtain support from abroad for conservation measures which the State could not otherwise be persuaded to adopt.

This illustrates another point. A treaty has the effect of calling attention of governments to the need to take conservation measures which they might otherwise consider unnecessary. A State is more likely to take measures to conserve a species or a habitat if other States have formally recognised the need to take such measures. A treaty sets an international standard below which a State may be reluctant to fall.

Are new international measures needed for habitat protection in Europe?
Scientists and conservationists need to do everything possible to protect the important remaining areas of lowland dry grassland throughout Europe. Within the EC, no new measures, except the proposed Habitats Directive and some conservation-based changes in the CAP seem to be necessary. But EC measures cannot apply in Central and Eastern European States which are not members of the Community. For them the only available international measures relevant to lowland dry grassland are the Bern and Bonn Conventions. The weaknesses of the Bern Convention have been mentioned above: there is also the question whether that Convention's close ties with the Council of Europe might discourage some Eastern European States from becoming parties to it.

There are at present no plans that I know of to write an agreement for lowland dry grassland species under Bonn Convention auspices. The question therefore arises whether effective conservation of lowland dry grassland areas in Europe, including the western republics in the U.S.S.R., needs a new treaty. It is worth mentioning that informal consideration has been given to the need for a new wildlife conservation treaty, which would not, of course, be limited to grassland by the staff of the U.N. Economic Commission for Europe. If a new treaty is needed, it could, in theory, be either an agreement under the Bonn Convention or a separate treaty.

This conference need not, and could not, resolve this question. It could however begin the process of visualising the essential features of a treaty on lowland grassland conservation, and discussing them with interested scientists in *all* the States which might ultimately become parties to it, to see whether the effort involved in preparing such a treaty would be likely to be worthwhile. In particular, it would be useful to begin by working out the essential features of a new international agreement involving the countries having the most important lowland dry grassland habitats, especially Spain, Turkey and the U.S.S.R. These are *not* the most important parties to the proposed agreement on white stork, so it is not clear how useful that agreement would be as a precedent.

The scientific material needed for a legal conservation case

The scientific material needed for a legal conservation case varies, of course, with the nature of the threat to the species or habitat and the legal basis (the clause in the treaty or in the national law) which is being relied on. But some general comments may be useful.

1. Conservationists must provide scientific arguments. The strength of a conservation case depends on scientific arguments and evidence. Conservationists should not leave it to others to collect and formulate the scientific arguments. Pressure is not an adequate substitute for a well documented case containing the kinds of materials described below. If conservationists do not provide it, either it will not be provided at all, or too little of it may be provided too late for an effective case to be made.

2. Making a legal case for conservation of a population or a habitat is an interdisciplinary task. This is so not merely in the sense that both a lawyer and an ornithologist are needed and must work closely together, but also that each must understand something of the other's discipline. In addition, depending on the nature of the conservation problem, it may be necessary to involve experts from other disciplines. For example, a case may need hydrological engineers to advise on the effects of lowering the water table, soil scientists and ecologists to advise on the effects of overgrazing (or reduced grazing) and erosion, phytosociologists to discuss the effects of seeding on vegetation composition and structure, entomologists to estimate effects on insect populations, etc.

It may also be necessary to have agricultural economists and sociologists to advise on the effects on the human communities of both the measures which constitute the threat (the economic effects of which may have been overestimated) and the measures proposed by the conservationists.

In future, it may be necessary to get the advice of a climatologist, when it is possible to estimate the effects of global warming in specific regions.

In general, we need to know more about the ecological effects of different low-intensity agricultural practices.

3. All this means that conservation organisations must have a professional approach. If the threat is a large scale one, the interests promoting it will be professionally advised and equipped. Fighting for a conservation case is not best done by amateurs, without professional experience.

4. In the cast of a threat to an area of scientific importance, conservationists should if possible have:
- published documented proof of the scientific importance of the area.
- published authoritative recognition of the scientific importance of the area.
- up to date quantitative information on the populations of the bird species using the area. (In the case of a staging post for migratory species, this should if possible include information on the turnover of birds during migration, and so on the total number of birds using the area, as distinct from the numbers present at any one time).
- ecological studies or surveys of the area which show what management is needed to maintain the scientific value of the area, how the ecosystem operates and what effects would be likely to result from the threatened measures.
- economic or agricultural information to counter any economic arguments which may be made for the proposed development.
- evidence that the area in question is an indispensable part of a network of reserves for the species or of the type of ecosystem in question. It should be possible to say whether there is any other area in the region to which the birds could go if displaced, and whether the other areas (if any) have space carrying capacity adequate for those populations.

5. In practice, the kinds of material needed cannot be adequately collected *after* the threat to the area has emerged. To make sure that it is available, it is essential to have regular bid counts and to carry out

ecological surveys and studies of the main features of the ecosystem *in advance*. "In advance" in most areas means "now". Studies take time. Conservationists have to convince universities and ecologists that the studies are worth doing on scientific grounds, not merely for future conservation purposes, however clearly foreseeable. So *all* areas of scientific importance need to be continuously monitored and actively studied if a strong case is to be made for their conservation when the need arises. As far as possible the ecological studies should investigate aspects of the habitat which are most likely to be directly affected by the foreseeable threats: lowering of the water table for irrigation, overgrazing, reseeding, as well as gross habitat change. Even if the threat which finally materialises is so gross that its destructive consequences are beyond dispute, ecological studies are likely to be valuable to explain why the area is scientifically important, why no other is comparable or no substitute is available, or why it would be difficult to recreate similar habitat elsewhere.

In addition, all available studies, including university students' papers, on all aspects of the area should be collected, even if they have not been written by scientists or with wildlife conservation problems in mind. All information is potentially useful. Information on e.g. farm yields, which may be crucially important often cannot be obtained after a controversy has begun: it may be too late to collect it, or those who have it may refuse to make it available.

Even if an area is primarily important for birds, *all* aspects of its scientific importance should be documented and used as arguments for conservation.

6. Conservation organisations in each country should get legal advice on the circumstances in which they have standing to bring legal proceedings against public authorities or developers, on different legal grounds. The necessary legal analysis should be done in advance, as far as possible, so that everyone concerned is familiar with it. This saves time.

7. Conservation organisations should identify, from among the scientists with most expertise on particular kinds of habitat, those who would make good witnesses in court. Sometimes scientists who initially gave clear opinions either change their views under questioning or qualify them so heavily as to leave only an impression of confusion and uncertainty. With this in mind, it would be useful to have an international list of suitable experts: sometimes a scientist from another country is free to express scientific opinion more forcefully than scientists working in the country where the controversy has arisen.

8. As far as possible, conservation organisations should use scientific material already accepted by the public authorities as authoritative. Courts dislike choosing between conflicting scientific opinions, as they do not feel competent to do so. This means that it is important to try to have carefully compiled lists of scientifically-important areas accepted officially *before* controversies arise.

9. Conservation bodies should learn to present their arguments clearly and carefully, not overstating their case, and saying precisely what conclusions should be drawn from each body of data. They should accept that there is often a margin of error in scientific opinions, and rely on the principle that in conservation matters (especially with species threatened with extinction) one must take the most prudent course of action. They must also identify very clearly the audience addressed by each argument, and not make the same arguments in the same way to politicians, courts, local farmers, and the media.

Influencing Agricultural policies for conservation: comments on David Baldock's paper

Baldock and Wells both rightly stress the importance of the current shift of agricultural policy away from intensification and the present radical reconsideration of the European Community's common agricultural policy. These give conservationists a better opportunity than they have ever had before to influence European agricultural policies in favour of nature conservation. In Central and Eastern European countries political and economic changes also provide new opportunities for trying to persuade the authorities to adopt policies consistent with nature conservation. It seems likely, although the question was not discussed, that the move from collective to privately owned farms will produce a variety of ecological changes, in particular as bank finance becomes available for mechanisation and fertilisation. It is not likely that all these changes will be desirable for nature conservation. All these developments need to be considered by conservationists not just in the context of different types of grassland or of birds, but of nature conservation generally.

Baldock's paper concentrates even more than Buxton's on the EC, and tries to describe the ecological effects of the Common Agricultural Policy (CAP) on grasslands. He rightly says that there is a lack of the most useful statistics. However, other means of getting relevant information, such as satellite photography, could be used, at least for some purposes.

Baldock identifies three influences for change: intensified use, drainage, and decline of livestock farming. The third is important where grazing is needed to maintain the habitat, and where reduced grazing is needed to maintain the habitat, and where reduced grazing gives rise to ecological succession away from grassland, or to deliberate choice of non-grassland uses. It is therefore important to identify areas where grassland is the natural climax vegetation, or at least a succession maintainable with minimal grazing or grazing by wild herbivores. On the other hand, intensified use of relatively infertile land can lead to overgrazing and deterioration of the vegetation, reducing the carrying capacity of the ecosystem for at least some bird species.

Where set-aside policies cause land to be left fallow, what is now grassland will remain so only in some areas of the Community. There is little point in encouraging set-aside if the effect would be ecological succession suppressing what remains of grassland habitat.

It is essential for clear thinking to distinguish between:
- measures to benefit or manage identified areas primarily for nature conservation.
- measures intended to promote conservation of types of habitat in areas which are not primarily dedicated to nature conservation but where the other uses can be partly or wholly reconciled with conservation objectives.

Areas to be managed primarily for nature conservation need:
- to be identified and listed, insofar as that has not already been done. This also means that the effective boundaries of the areas need to be defined.
  public authorities responsible for nature conservation (at local, regional, national, European Community or international level) should be persuaded to accept officially the designation of these areas.
- management policies or plans for each of these areas need to be prepared. This does not need to await the official recognition of the importance of an area. It can and should be done, by either official or private bodies, even before the area is officially recognised as one which should be managed primarily for nature conservation.
- the management policy or plan should be put into operation.

As Baldock points out, France and Spain together contain a large proportion of the lowland dry grassland in the Community. The most important of these areas for birds in France and Spain have been listed in Grimmett and Jones (1989). Some at least of these areas are extensively managed semi-natural habitat which is probably the most important scientifically and also relatively easy to conserve if the political will can be found. In effect, these areas are worth making into nature reserves and relatively easily manageable for that purpose. As far as areas important for birds are concerned within the European Community, the authoritative inventory, called for by Baldock, already exists. What may *not* yet exist is an adequate series of management plans for the ideas identified by Grimmett and Jones (1989(. What certainly does not exist are socio-economic plans offering local people a profitable means of using the land which involves one minimum social change or inconvenience which is compatible with conservation. Such plans are needed to persuade local people to accept whatever ecologically-orientated management is needed.

In addition to the special environment-based measures mentioned by Baldock (environmentally sensitive areas, involving modified farm practices, and set-aside, involving cessation of farming) one should mention arrangements to ensure that no EC funds are spent on changes in areas of scientific importance listed by the EC Commission for the purposes of the Birds Directive. It seems unlikely that modifications in the CAP itself would go so far as to create incentives for useful habitat creation measures: habitat creation, to be scientifically valuable, would need to be done under the ACE (Action by the Community relating to the Environment) scheme.

Measures intended to benefit specific areas primarily intended for nature conservation, or areas or parts of wider areas subject to multiple use are both necessary (it will never be possible to have enough nature reserves, or large enough nature reserves, to conserve all European species primarily or exclusively in reserves). The two kinds of measure need to be integrated with one another. In particular, in areas where grazing pressure from domestic animals is needed to maintain grassland and prevent vegetational succession, precisely judged management, based on appropriate funding or other measures, may be needed. But of course such measures need not be adopted at Community level: they could be national or regional measures.

Areas that are not being managed primarily for nature conservation purposes may nevertheless be influenced favourably either by region-specific measures such as Environmentally Sensitive Area arrangements or regional set-aside provisions, or by the effects of general agricultural policy measures such as price-supports. "Extensification" and ESA measures can vary widely in content and in the degree of strictness and therefore of protection given, in practice, to particular habitat types. In thinking about the effects of all such region-specific measures it is essential to be clear whether one is talking about measures applying only to parts of (relatively) large farms or about measures applying to whole farms in whole localities. Measures designed only to persuade or pay farmers to set aside, or to manage in some special way, parts of their farms are unlikely to produce or conserve areas large enough to be valuable to many endangered grassland species. (Corncrakes however, could probably benefit significantly from a patchwork of relatively small areas).

In discussing all such measures it is essential to be as clear as possible precisely what one is trying to achieve, and for which species, and how the steps proposed are likely to affect the incomes of the local people. It is also essential to remember that all regimes under which money is given on condition that certain policies are carried out by farmers need supervision to ensure that the conditions are respected, and this causes administrative costs and inconvenience which is unlikely to make the measures popular. The more precise the results, in habitat conservation terms, which are sought, the more supervision will be needed, and the greater the risk that the whole scheme will become an unpopular nuisance in which farmers are unwilling to cooperate. Also, there is always a risk that public opinion may come to think it unsound to pay farmers not to use their land, especially when other public money is being used to support farm prices.

Within the European Community, the Common Agricultural Policy (CAP) is now being radically reviewed. Whatever changes are ultimately made will almost certainly have considerable ecological significance, especially if they result in marginal land being taken out of agricultural production and either abandoned or used for forestry or other purposes. Reducing the level of price supports will, in itself, tend to take less productive land out of production. However, the Commissioner for Agriculture has suggested measures designed to support large farms less and small farms more, relatively, than at present. Small farms are not necessarily uneconomic, nor are they always on marginal land, although some are both. To the extent to which small farms, in some regions of the Community, become uneconomic due to reduced prices (not fully offset by preferential measures in favour of small farms), it is marginal land which is likely to become available for other purposes. Also, prices for farmland in general are likely to decline, so that conservation interests will have to pay less to buy land for wildlife purposes.

For conservation it is important that preferential measures in favour of small farms should not be in the form of e.g. payments per head of livestock, or other measures which tend to promote over-stocking and over-grazing, or other environmental damage.

Even without excessive detail or bureaucratic control, there is considerable scope for more conservation-orientated use of set aside and ESA measures, for example to create buffer zones around nature reserves, and to protect areas listed as potential nature reserves.

ESA and similar arrangements probably work best if they operate through standard management agreements voluntarily accepted by local farmers. It would be difficult, and probably counter-productive and unwise, to try to impose positive obligations on farmers to farm in a particular way.

FURTHER RESEARCH AND WIDER IMPLICATIONS

Ornithological research is urgently needed to estimate the present populations of a number of lowland dry grassland bird species, and to discover whether they are under immediate threat (Tucker, Rodriguez and Juana). Standardised census methods should ideally be used, and long-term monitoring of as many species as possible should be planned, in particular where threats seem most acute, and to see whether conservation measures have been effective. The possibility must be kept in mind that small, perhaps inconspicuous species may prove to be better indicators of the quality of dry grassland ecosystems than the larger, more striking birds with slower reproduction rates, smaller populations, and larger territories. Because of the scale of the project and the urgency of the situation, research on the bird populations of the area of SE Anatolia likely to be flooded as a result of the Ataturk dam needs special attention (Baris).

Other kinds of research which are needed are plainly inter-disciplinary and ecological, and would not be carried out only by ornithologists.

Several speakers pointed out that research is needed into grazing pressures. Ideally we need to know what number of livestock of appropriate species and breeds (or of wild herbivores) is needed in each kind of habitat to maintain vegetation suitable for each grassland species likely to nest there and, ultimately, for each species of flora also. Appropriate experiments would not be difficult or expensive to carry out, if suitable land was available for them, but it must be remembered that different grazing *methods* produce different results, and that the same grazing pressure and method will produce different results in different soil and climatic conditions (Hopkins).

Further research is also needed to make optimum use of land which is going out of agricultural use (or going out of intensive agricultural use) as a result of set-aside or similar policies. What kinds of areas should be chosen for restoration or re-establishment of grassland habitats? Intuitively one assumes that areas near any remaining undamaged sites should be chosen, provided that they have similar soils and are otherwise ecologically comparable to facilitate natural recolonisation by both birds, flora and invertebrates. But the reality may be more complex. Ground that has been intensively farmed may currently have low productivity but may not revert to floristic richness, even if propagules are available, without special management.

Where ESAs are going out of intensive use and into "extensive", i.e. low-intensity use, are nature reserves needed to protect local seed banks for recolonisation of the areas concerned? If no reserves or undamaged ecosystems exist locally, can similar results be achieved by restoration or re-establishment in a small central area, from which recolonisation will then take place? How far can the restoration techniques described by Wells be used in other parts of Europe? In practice, how far should agricultural policy measures go in specifying precisely what farming techniques should be used to produce the optimum conditions for conservation? One cannot assume that "traditional" farming methods will continue entirely unchanged in ESAs, or that they will be precisely adapted to the needs of threatened species in conditions of post-intensive use. The answers will be different in different regions of Europe, but this may mean that research is needed in each region to obtain satisfactory answers.

Several species (great bustard, stone curlew, demoiselle crane *Anthropoides virgo*) which are traditionally regarded as grassland nesters now seem to prefer arable land. It would be valuable to discover why this is so. Has the quality of the available grassland deteriorated, or only its attractiveness early in the season when nest sites are chosen, and if so, why? Is the arable land producing more invertebrates or other food, early in the season, than the grassland, due to higher nutrient levels or different thermal conditions? If neither the vegetation or vegetation structure in the grassland has altered, has its invertebrate fauna declined? Or, as Kollar suggests, did these species in the past use mixed areas of ploughed fields and fallow grassland, and is the reduced breeding success on modern ploughed land due to the use of machinery? Again, it would not be difficult to devise appropriate experiments or surveys if land was available to carry them out.

Some contributors (Tucker, Flint, Bignal) mentioned the problem of species which choose arable areas for nesting early in the season, only to find later that nesting and chick survival is low as a result of harvesting or intensive grazing. Research is needed into methods of making any nearby areas of untilled habitat more attractive, as well as means of reducing the dangers from machinery. Even if the techniques used in the USSR of taking eggs for artificial incubation were widely acceptable, they are clearly impracticable with most species smaller or more numerous than great bustard and steppe eagle *Aquila rapax*. The Austrian technique of providing weed-rich areas as feeding areas for bustard chicks (Kollar) may also be difficult on a large scale.

In Central and Eastern Europe, the ecological effects of current changes in agricultural practices need to be carefully watched. Collective farms were apparently more damaging to nature conservation than private farms, presumably because they were more mechanised and used more fertilisers and chemicals. If this is correct, as already mentioned, one would expect private farms to become more damaging as finances become available for more intensive agriculture or, as is already happening, private farmers turn to horticulture, which offers a cash crop saleable directly rather than through remaining official distribution channels.

It will be seen that the challenge of conserving grassland birds, almost all of which are dispersed species most of the year, is not only to devise workable conservation policies applying to large areas subject to farming or other non-conservation uses, but also to integrate nature reserves and socio-economic policies for human welfare in these areas. This is a much more demanding task, and a much more multidisciplinary one, than working out management policies in areas where conservation has priority. But it is also a more worthwhile task. It will oblige conservationists to develop expertise in other disciplines, or to work closely with economists and agriculturalists who already have skills in other disciplines. It will oblige conservationists not merely to lobby and to sign petitions, but to work out quite complex policies to a highly professional degree. It will not be enough, either in western Europe or in the former communist countries, simply to urge the authorities to achieve a particular result, leaving it to them to discover how to achieve it and to operate the agricultural and other policies needed to bring it about.

Because farm practices, and social and economic circumstances, differ from country to country, the agricultural policies which are needed will have to be adapted to each State, perhaps even to each region. No single formula would suit the whole of Europe, or the whole of the European Community. In the case of ESAs and similar measures, much will depend on the willingness and ability of the relevant authorities, national, regional or local, to work out and supervise conservation-orientated regimes. Set-aside is basically a simple concept, but if it were adapted to achieve precise objectives it would have to involve some more detailed rules and some supervision. But, as Baldock said, once an authority is paying an annual sum of money per hectare, it can impose environmental requirements, simple or detailed, general or specific. It is for conservationists in each country to work out carefully the minimum number of relatively simple measures needed to achieve the most urgently needed results, and to persuade the appropriate authorities to adopt them. The aim should be to get a basic framework set up and operating: details, and improvements, can be added later, in particular in the light of more precise ecological knowledge about the needs of particular species, and experience in the working of the measures in question. Important as the research outlined here undoubtedly is, it must not be used as an excuse for doing nothing until it is completed. The seminar not only showed that the situation of many grassland bird species is serious, it also showed that we already know enough about what is needed to get started on conservation measures.

If an ornithologist may be permitted to say so, botanists need to produce as quickly as possible authoritative inventories of sites of botanical importance throughout Europe, at least of those sites which are important enough to be potential nature reserves or ESAs or to require conservation otherwise using standard criteria (Van Dijk; Géhu 1984). The flora of European grassland cannot be satisfactorily conserved only in nature reserves, but it cannot be conserved without them either.

Some research into the ecological needs of grassland birds in their wintering areas, at staging posts on migration, is also needed. Although we probably do not need to look beyond breeding areas for the causes of the decline of grassland species, populations can be as seriously affected by factors influencing wintering areas as by factors operating in nesting areas.

Several dry grassland bird species occur in areas of poor fertility. Several speakers mentioned that low fertility is necessary for certain plants (and probably also for certain invertebrates). When areas are being selected for conservation purposes, areas of appropriately infertile soils should be chosen for such species. It is not enough to look only at present vegetation, to decide how easily management can maintain or re-establish short, sparse vegetation. Also, areas of low fertility soils are cheaper to buy, or if managed, will involve less conflict with farming interests, than areas of inherently higher fertility.

Although many dry grassland birds are partly or wholly insectivorous, at least at some stage in their life cycle, there was no paper at this seminar on dry grassland invertebrates, and we do not know enough about their populations or ecology for management purposes. This is a huge field for research, but a necessary one. Re-establishment or maintenance of plant communities of floristic richness and appropriate vegetation structure will probably produce good invertebrate populations automatically, but we cannot be certain that it will do so. Once again the need for interdisciplinary studies involving both ornithologists and others is obvious. The first step would be faecal analyses of the insect food of larger bird species.

Even in areas of intensive agriculture, pockets of less damaged vegetation often persist in places inaccessible to ploughs and grazing livestock, such as cliffs and river banks. These pockets can be used, in the absence of formal exclosure experiments, to see what a relatively undisturbed flora is like in the region, as this could be a source of seed. Most such pockets are, however, too small to hold an adequate range of invertebrates, so re-establishment of grassland based on such pockets would not be likely to restore the original diversity.

Apart from the need for a comprehensive inventory in regions for which this conference, at least, has incomplete information, other questions arise:

(a) Some lowland dry grassland areas, especially in the Union Soviet Socialist Republic, are used by important species such as great bustard only during part of the year. Where are the wintering areas of these populations, and do they need conservation?

(b) In most of western Europe, lowland dry grassland areas are discontinuous and fragmented. How much gene flow is there between these areas? If the population of a bird species died out in such an area, would it be recolonised naturally by birds from other similar areas? Should we be concerned about the potential effects of inbreeding in very small discrete populations? No doubt the answers are different for different areas and for different species.

(c) Which areas of lowland dry grassland represent the natural climax vegetation in the region, and which could be maintained only by ensuring e.g. some level of grazing other than that provided by indigenous wild herbivores? Can we get information useful for bird conservation by exclosure experiments or some other fairly simple ecological research, or do we need to ask for large scale research help from phytosociologists?

(d) Which bird species (if any) can be satisfactorily conserved in western Europe only in nature reserves, even assuming that all or most of the remaining areas of lowland dry grassland could be made into reserves, and which species will need, as well, multiple use policies covering much larger areas? Will even a large network of reserves create isolated populations in the case of sedentary species, and would such populations be vulnerable to inbreeding or to stochastic events, as mentioned above? Even if populations are separate most of the time, is there interrupted gene flow resulting from e.g. hard winters bringing otherwise separate populations into contact?

(e) What additional conservation problems can be foreseen as a result of global warming? Some lowland dry grassland habitats would probably be seriously altered by even relatively small increases in temperature or decreases in precipitation. At least some computer simulations of the regional effects of global warming suggest considerable increases in temperature and reduction in precipitation in the steppe areas of the south-west of the USSR. Even a relatively modest expansion of the Trans-Caspian desert might have very serious effects on the remaining areas of protected steppe habitat in this region.

(f) Apart from the effects of global warming, do we need to measure and monitor gradual changes in the quality of lowland dry grassland and in its suitability as habitat for the most important bird species involved? Changes due to e.g. overgrazing might easily occur without being noticed (or might be noticed too late) without monitoring. Efforts should certainly be made to identify areas where grazing pressure has increased significantly in recent years, whether due to the CAP, other official policies or otherwise. It cannot be assumed that in e.g. Turkey over-grazing has gone on for so long that steppe birds have adjusted to it, or benefit from it, and that it cannot constitute a threat to the survival of current population levels. These questions may be important because habitat deterioration of relatively subtle kinds may be contributing to the decline of e.g. lesser kestrel. Undetected declines could easily occur, or be occurring already, in populations of smaller, less conspicuous, or less easily counted species. One should also be aware of the possibility of changes in grazing pressure due to e.g. changes in the extent, or the patterns, of *transhumance* (altitudinal movements of domestic grazing animals in spring and autumn).

CONCLUSIONS

The main problems of conservation of habitats and species *outside* nature reserves, discussed or at least raised in this seminar, are not confined to grasslands or to grassland birds. Conservation of dispersed species of both fauna and flora will oblige conservationists to make the maximum use of existing and future legal instruments, and to try to influence official agricultural, forestry and physical planning policies into approaches compatible with, or promoting, nature conservation. To do all this conservationists in western Europe need to be more professional, better prepared, and more interdisciplinary in their approach than in the past. In central European countries conservationists also need to be more activist in future, and to take more initiatives, than has been possible until recently. There is much scope for conservationists to learn from one another's experience in conservation controversies, if cooperation is close and information efficiently exchanged. There is not enough cooperation of this kind between ornithologists in different countries, or between ornithologists and scientists in other disciplines. These kinds of cooperation need to be deliberately promoted, in addition to any cooperation which may come about for the purpose of drafting new international agreements.

In summary, conservation organisations, on an interdisciplinary basis and with professional standards:

1. When proposing new or modified national or international conservation measures, should be clear precisely why they are needed and what aim is to be achieved, and how.

2. If they wish to see international conventions implemented more effectively, should draft the necessary supplementary resolutions and Agreements.

3. Should get legal advice as to their rights to challenge environmental threats in national courts, especially those threats which are inconsistent with national obligations under European Community law and under Conventions.

4. Should not be content with lobbying, but should prepare scientifically well-documented cases for conservation of specific areas, *before* threats arise. This necessitates maintaining up to date bird counts and ecological surveys.

5. If they wish to influence general EC agricultural measures in favour of nature conservation, should carefully formulate precise and fully-argued proposals and think through their precise implications.

6. Should seek arrangements to ensure that no public funds are available for developments affecting areas listed authoritatively as of international scientific importance.

7. Should remember that the most important areas of lowland dry grassland in Europe are in Spain, Turkey and the USSR, and should give these countries priority in their conservation efforts.

8. Should themselves promote and carry out research into all important aspects of the management of scientifically important areas, with a view to preparing management plans and if necessary new international measures.

REFERENCES

Batten, L. (1987). The effectiveness of European agreements for wader conservation. *In:* Davidson, N.C. and Pienkowski, M.W. (eds). *The Conservation of International Flyway Populations of Waders*. Special Publication 7: 118-121. *Wader Study Group Bull.* **49**, International Waterfowl Research Bureau.

Curry-Lindahl, K. (1964). *Europe: a natural history*. Hamilton, London.

Géhu, J.M. (ed.) (1984). *La végétation des pelouses calcaires. Strasbourg 1982*. Cramer, Vaduz.

Goriup, P. and Batten, L. (1990). The conservation of steppic birds: a European perspective. *Oryx* **24**: 215-223.

Grimmett, R.F.A. and Jones, T.A. (1989). *Important Bird Areas in Europe*. ICBP Technical Publication No. 9 International Council for Bird Preservation, Cambridge.

IUCN (1990). *The lowland grasslands of Eastern Europe, A Survey, with selected country case studies*. IUCN, East-European Programme.

Knystautas, A. (1987). *The Natural History of the USSR*. Century, London.

Temple Lang, J. (1982). The European Community Directive on Bird Conservation. *Biological Conservation* **22**: 11-25.

Temple Lang, J. (1990). International and Legal Aspects of Conservation of Irish Bogs. *In:* Schouten and Nooren (eds.), *Peatlands, Economy and Conservation*. SPB Academic Publishing, The Hague.

THE STATUS OF SEMI-NATURAL GRASSLANDS IN EUROPE

Gerard van Dijk[*]

Roerdompstraat 31, 3815 BT Amersfoort, Netherlands

ABSTRACT

Semi-natural grasslands were very widespread in Europe at the turn of the century. Though they are mostly secondary formations, they represent an enormous genetic resource of both flora and fauna. However, the greater part of this resource has been destroyed in the last 50 years. Not only has the area dramatically declined, but quality has also suffered due to factors such as fragmentation, improper management and acid, nutrient-rich precipitation. In several countries, serious efforts are being made to conserve the last relics and to restore lost grassland vegetation. Nevertheless, further loss of genetic material is inevitable and attempts to regenerate vegetation will suffer from demographic problems (absence of species) and because the genetic basis of future populations has been narrowed.

This all makes conservation of the remaining species-rich grasslands very urgent. International attention should be aimed primarily at the maintenance of large genetic reservoirs, especially those lacking adequate protection so far. This includes calcareous grasslands and river valleys in France, rough grazings in Great Britain, parts of Ireland, large areas partly in river valleys in some east European countries (e.g. Poland, Yugoslavia) and grasslands in the USSR, about which, however, few data were available. Although little information is available, it is clear that large areas of grasslands in Italy, Spain and Portugal have been or are being converted to arable land or forest.

The main instruments for conservation are: creation of nature reserves and proper management if the gap with present land use is too large, introduction of management agreements where agriculture is still extensive, and stimulation of low-input agriculture in areas to be protected. In addition, afforestation plans and any agricultural development (land consolidation, drainage, regulation of rivers) must be critically evaluated before being allowed to proceed.

INTRODUCTION

In the last one hundred years of nature conservation, important efforts have been made for the protection of threatened habitats and species. Much attention, though alas not always successful, has been given to the protection of natural ecosystems and those semi-natural ecosystems that are no longer of economic value, like heathlands and marshes. However, though they were amongst the first to suffer greatly from modern technologies, grassland and arable land ("pseudo-steppe") vegetation has been relatively neglected, except in those cases where grasslands were of great ornithological importance, like in the Netherlands.

Though a serious decline of grassland bird populations is still taking place, it is less serious than the massive decline of semi-natural vegetation and related invertebrate fauna. Protection measures for breeding waders take place on a larger scale and have better prospects. The situation for wintering species, like geese, even justifies optimism as populations have increased considerably during recent decades. Moreover, birds often fare well in fertilized grasslands.

The contrary is true for most plants, which suffer increased competition by grasses once a level of some tens of kilograms of nitrogen per year per ha is reached (Oomes 1983, Delpech 1975). This level of fertilization was reached in the Netherlands about half a century ago. Of course many other major changes in land use have contributed greatly to the loss of species-rich grasslands, like ploughing (followed by reseeding or conversion to arable), drainage, and, especially in recent years,

---

[*]Staff Officer, Ministry of Agriculture, Nature Management and Fisheries; recommendations here are made in a private capacity

abandonment. However, none of these seems to have had such an overall effect as the use of artificial fertilisers. This explains why the loss of semi-natural grasslands, even more serious in scale than in other habitats, could take place almost unnoticed. Excellent information on drainage in four EC countries is given by Baldock et al. (1984).

Even amongst conservationists, semi-natural grasslands are not always sufficiently appreciated, or their need for proper management is hardly recognized. In the Netherlands, the change in view came about 50 years ago, greatly stimulated by V. Westhoff. In recent years, however, different "schools" of thinking have developed, varying from adhering to traditional management (including mowing) to a preference for more natural processes such as grazing, though cattle densities may be higher than in natural situations.

In this context it must be mentioned that not all ecologists still believe that forests were formerly dominant everywhere in western and central Europe. Ellenberg (1986) states that grasslands in central Europe are anthropogenic. F. Vera (pers. comm.) supposes that the proportion of open land may have been up to some tens of per cent of the total. Geiser (1983) gives several arguments why it is probable that open areas covered a large proportion of the land under natural conditions. Though this discussion may go on for some time, knowledge about this is important for the general appreciation of grasslands and for formulating management strategies.

In some countries, where important strongholds of semi-natural grasslands remain, few efforts have been made for the protection of grasslands until recently. This applies for instance to France, where destruction is now very fast, and Poland, where considerable efforts for nature protection have been made, but little for grasslands. Even if species-rich grasslands are part of reserves or national parks, the need for management seems not to be acknowledged.

The aim of the present contribution is to stimulate conservation measures, especially in those areas where there is a combination of important biological features and a lack of adequate protection.

THE MAIN TYPES OF SEMI-NATURAL GRASSLANDS

Speaking about semi-natural grasslands implicitly excludes natural grasslands. These are, in western and central Europe, mainly alpine vegetation, summits of secondary mountain chains, salt marshes and sand dunes. In the past, other types of natural grasslands may have been rather common. Some seemingly secondary grasslands, like the calcareous grasslands of the Somme Valley in France (Stott 1981) are thought to have persisted even through one or more glaciations, and others at least through the post-glacial period (Yorkshire Wolds; Bush and Flenley 1987). Today, however, the great majority of lowland grasslands are certainly anthropogenic, although they may serve to some degree as substitutes for original grasslands, once maintained by wild grazers and browsers.

The main types of semi-natural grasslands in northwest Europe are here described according to the Braun-Blanquet system. This system is widely used on the continent, but is also found in recent publications about Great Britain (Shimwell 1971, Willems 1980, Riely and Page 1990), Ireland (O'Sullivan 1982) and Scandinavia (e.g. Losvik 1988). Concise information about the botanical composition of the units of the different levels of this hierarchical system is given for central Europe by Ellenberg (1986) whereas Riely and Page (1990) give a good survey of British vegetation according to the same system. Westhoff and Den Held (1968) give very detailed information about the vegetation occurring in the Netherlands (and often also in adjacent countries). Therefore it is not considered necessary to detail here the characteristic species of the units involved.

Nevertheless it must be stressed that practically all the types mentioned are of great importance for the protection of an often great number of species. In their diversity they hold a great variety of species of plants and animals. In calcareous grasslands, for example Willems (1990) mentions about 700 different plant species, including some 200 bryophytes and lichens; on less than 1 ha in the *Kaiserstuhl* there were 56 butterfly species and 131 species of bees. According to Holzner et al. (1986), 1,041 species of insects depend on Austrian dry grasslands, of which 85 per cent are Red List species. The same applies to 150 Red List species of beetles. On a dry grassland area of less than 10 ha 1,080 species of butterflies and moths were recorded. Korneck and Sukopp (1988) list 588 species of higher plants for dry grasslands and 297 for wet grasslands.

Not only is the conservation of all these types necessary, but also all the variations within these types, as only this can guarantee the survival of the individual species concerned in sufficiently large populations. Not only do demographic factors play a role, but also genetic ones. The loss of genetic material, due to populations becoming too small, is a serious problem (Ouborg 1988).

The following brief survey only gives the names of the different units, based on the above-mentioned literature and some further information (Figure 1). In the text of this paper, units of different levels will be used, according to the information available. For example, all calcareous grasslands can be referred to as the Festuco-Brometea class. In that class, two orders are distinguished, one for eastern and one for western Europe and they comprise together five alliances, which can be further divided into more associations. For each level there are characteristic species.

The main types are given in the following order: class (ending "..*etea*"), order ("..*etalia*"), alliance ("..*ion*"). Associations ("..*etum*") are not listed, but will be mentioned in the text if special information is available. Sometimes there may be some departure from certain views, as in the case of Cynosurion, the species-rich part of which is placed in Arrhenatherion and the rest in Agropyro-Rumicion crispi by Westhoff and Den Held (1968). Here Ellenberg (1986) is followed.

The scope of this survey is limited to western and central Europe. For some countries other alliances, not listed in this survey, are mentioned in the text.

In the Mediterranean part of France, seven other alliances occur, and four more at higher altitudes (R. Delpech, pers. comm.). No doubt even more alliances occur in other southern regions.

THE STATUS OF SEMI-NATURAL GRASSLANDS BY COUNTRY

The information in this survey about different countries is very heterogeneous. Though it is clear that much more information exists, the time available to collect this was too limited to get a more complete picture. Nevertheless, it is clear enough that in most countries semi-natural grasslands are under great pressure.

The Netherlands

The total grassland area is 1.1 million ha, more than 50 per cent of the agriculturally used area (2.0 million ha). Both fertilization and organic manuring are likely to attain unique levels in Europe, if not in the world. This is the main reason for the enormous decline of Dutch semi-natural grasslands.

In 1986 the average application for the whole area in agricultural use was 249 kg N per ha per year as fertiliser and 139 kg N per ha per year as manure

---

1	**Molinio-Arrhenatheretea: so called-neutral grasslands** (Riely and Page 1990). (These may be better referred to as "moist to moderately dry", Ellenberg, pers. comm)	**3**	**Sedo-Scleranthetea: arid grasslands of rocks and sand**
1.1	Molinietalia: periodically wet meadows		Koelerio-Phleion phleoides (formerly: Armerion elongatae, in Rothmaler 1981 in the Class Sedo-Scleranthetea): acid semi-dry grasslands. Though Ellenberg (1986) placed this alliance in 2 (above), he now includes it here (Ellenberg, pers. comm.).
1.1.1	Molinion: meadows on moist, poor soils (but rich in species)		
1.1.2	Filipendulion: vegetation of high forbs (develops e.g. in case of abandonment of Calthion)		
1.1.3	Calthion: meadows on moist, rather fertile soils		Further types mainly occur in areas without agricultural use, e.g. dunes, and will not be dealt with here.
1.1.4	Cnidion: riverine meadows in eastern central Europe		
1.1.5	Juncion acutiflori: moist meadows rich in rushes	**4**	**Nardo-Callunetea: heathland and acid grasslands on very poor soils**
1.2	Arrhenatheretalia: damp and moderately dry meadows		
1.2.1	Arrhenatherion: hay meadows of fertile, naturally well drained soils		In the British Isles acid grasslands are still in agricultural use, whereas on the continent they have become extremely rare. Hence, they will be included here.
1.2.2	Polygono-Trisetion: as the latter, but at higher altitudes		
1.2.3	Cynosurion: replaces 1.2.1 in case of regular grazing	4.1	Nardetalia: acid grasslands
1.2.3	Poion alpinae: pastures of higher altitudes (but still anthropogenic)	4.1.1	Eu-Nardion: mountainous acid grasslands
		4.1.2	Violion caninae (= Nardo-Galion): acid grasslands of lower altitudes. This is the same alliance, as Violo-Nardion, mentioned by Ellenberg (1986).
2	**Festuco-Brometea: dry calcareous grasslands (on chalk and limestone)**	4.1.3	Juncion squarrosi: seems related to 4.1.2; no further information
2.1	Festucetalia valesiacae: continental dry and semi-dry calcareous grasslands	4.2	Vaccinio-Genistetalia: heathlands
2.1.1	Festucion valesiacae: continental dry calcareous grasslands	**5**	**Violetea calaminariae: heavy-metal vegetation**
2.1.2	Cirsio-Brachypodion: continental semi-dry calcareous grasslands		The communities are open or closed and they mainly consist of hemicryptophytes. They often occur in mosaics with Koelerio-Corynephoretea, Festuco-Brometea, Arrhenatherion and Nardo-Callunetea (Westhoff and
2.2	Brometalia erecti		
2.2.1	Xerobromion: suboceanic extremely dry calcareous grasslands, perhaps partly natural		
2.2.2	Mesobromion: suboceanic semi-dry calcareous grasslands		

Figure 1: Types of semi-natural grasslands in Europe considered by the present survey

or slurry, making an annual total of 388 kg N per ha (Centraal Bureau voor de Statistiek 1989). About 50 kg N per ha per year from precipitation can also be added. This total is some ten times more than semi-natural vegetation can withstand. The high amount of organic manure or slurry is not only due to high stocking rates, made possible by fertilization, but also to the importation of fodder for 8.6 million pigs, 74 million chickens and as a supplementary feed for over 2 million (mainly dairy) cows (Centraal Bureau voor de Statistiek 1989). The total number of cattle was 4.7 million (Logemann, pers. comm.).

In 1988 estimates were made of the remaining semi-natural grasslands (and other habitats) in the Netherlands (Bakker *et al.* 1989). The dry grassland category included about 10,000 ha (20,000 ha dunes excluded), and wet grasslands 9,000 ha. The total, 19,000 ha, is 1.7 per cent of the national grassland area. The authors add that these numbers are too optimistic, for the truly species-rich area is (far) lower.

Dry semi-natural grasslands (Arrhenatherion and Mesobromion) in river floodplains for example, cover 2,530 ha in the statistics, but the net surface is known to be about 500 ha. True *Cirsio-Molinietum*, which is rich in rare species, covers only about 100 ha, probably far less than 1 per cent of the former area (which may have been hundreds of thousands of hectares, according to Sissingh 1978). The scattered remnants suffer much more from external hydrological influences than dry grassland habitats. Calcareous grasslands on rendzina soils cover at most 30 ha.

A national survey of threatened and extinct plants (Westhoff and Weeda 1984) listed 440 species, of which 94 were grassland plants. The authors say dry grasslands have suffered most, and from fertilization in the first instance. In an earlier survey, Westhoff found that 170 out of 400 rare species were really protected in nature reserves, only 28 of which were grassland plants.

The above-mentioned 500 ha of grassland in river floodplains cover 1.5 to 2 per cent of the Rhine floodplain system. They comprise *Arrhenatheretum*, *Lolio-Cynosuretum* (species-rich forms of Arrhenatherion and Cynosurion), *Medicagini-Avenetum* (Mesobromion) and mixtures of these communities. They cover the best drained parts of the floodplains and may formerly have covered roughly 50 per cent of them. The decline is about 95 per cent or more. The main cause is use of fertilisers, followed by clay and sand extraction and dike reconstruction.

In the south of the country 20 ha of calcareous grassland (Mesobromion) remains in nature reserves, scattered in 20 sites of 0.2-4.0 ha (Willems 1990). This may be only 5 per cent of the original area (Willems 1987), but they are extremely important as genetic reservoirs, with 200 vascular plant species and 120 bryophytes and lichens and large numbers of fungi, ants, bees and wasps (Willems 1990).

In the province of Friesland only 3,500 ha, about 1.5 per cent of the province, is unfertilized grassland. Yet 60 per cent of the terrestrial plants are more or less confined to this habitat (Schotsman 1988).

Though no less than 235,000 ha is managed by the state forestry service and private nature conservation bodies, only a small part consists of semi-natural grasslands as dealt with in this paper.

Recently a new Nature Policy Plan has been adopted, as a framework for future nature conservation and "nature development" measures. Eventually, a total of 200,000 ha of farmland (about 10 per cent of the total), mainly grasslands, will be given special protection, half by management agreements and half by acquisition (and, pending that, management agreements). Given the intensity of land use in the Netherlands, the only possibility for regeneration of grasslands, which will be an important issue in several parts of this area, is acquisition, followed by stopping the use of fertilisers. Management agreements are, however, useful for so-called meadow bird management and for vegetation along ditches (strips of 3-10 m width).

In 1990, management plans were ready for 50,000 ha. An area of 13,000 ha was covered by different sorts of agreements and 8,600 ha of the 100,000 ha targeted had been bought since 1976.

Another 50,000 ha will be bought for "nature development", which mostly means the creation of other habitats, like wetlands, but which may comprise semi-natural grasslands as well.

An area of about 50,000 ha consists of linear elements like roadside verges and railway banks (Westhoff and Weeda 1984) and thanks to the efforts of P. Zonderwijk the majority are now managed without the use of fertiliser and herbicides. Though almost half of the Dutch flora is found there (Westhoff and Weeda 1984), these linear sites are not an adequate substitute for proper nature reserves (K. V. Sykora, pers. comm.).

Much of the success will depend on the possibility of safeguarding the remaining relics of semi-natural grassland ("hot spots"), before they have disappeared. Different investigations have shown that the seed bank in the soil is a very serious limiting factor for the restoration of grasslands (Bakker 1982, 1985; Sykora and Zonderwijk 1986; Beltman, pers. comm.), so that we are very dependent on relict populations. Migration is often a problem, too, but even if it were not, there are no longer large genetic reservoirs, neither in the country, nor across the border.

Great Britain

The great majority of land-use in Britain is agricultural. For the UK (including Northern Ireland) this amounts to 76 per cent of the total land

Plate 1: *Fritillaria* rich grassland along the river Vecht, Netherlands [photo: G van Dijk]

Plate 2: Rough grazing along a valley in Scotland [photo: G van Dijk]

area (Lee 1990). In fact, large areas would be considered as "natural" in other countries, but still have an agricultural use. Large areas of rough grazings and moorlands persist, whereas on the continent most such habitats have been reclaimed or afforested.

According to Duffey et al. (1974), of the c. 1,500 species of flowering plants in Britain, over 500 are associated with grasslands and about 400 are most frequent in this habitat.

Ratcliffe (1984) estimates that only 5 per cent of permanent lowland grasslands now remain agriculturally unimproved neutral grassland. He describes the decline of plant communities and their related species. He further estimates that no more than 20 per cent of the pre-war chalk grassland area survives with its former floristic richness.

Lowland grasslands
The total area of lowland grassland in England and Wales declined from 7.8 million ha 50 years ago to 4.8 million ha now. Between 1939 and 1945 40 per cent disappeared through ploughing (Fuller 1987). The definition of "lowland" was not purely topographical, but the majority of the land included was under 240 m above sea level. Of this lowland grassland only 0.6 million ha (11 per cent) could be considered semi-natural, of which only 0.2 million ha is left if we exclude the rough grasslands (i.e. only 3 per cent of its area 50 years ago, Fuller 1987). The decline of unimproved and rough grasslands together from 1932 on was 92 per cent. However, the decline in rough grasslands alone was "only" 65 per cent.

Apart from ploughing, fertilizing was an important cause of change. In 1938 almost no grassland received nitrogenous fertiliser, in 1944 15 per cent; in 1960, 28 per cent; in 1976, 76 per cent and in 1985, 85 per cent (Fuller 1987).

In the period 1939-1971 unimproved permanent lowland grassland declined by 70 per cent. After 1970, ploughing (followed by reseeding) was no longer considered so necessary.

Fuller supposes that perhaps only a small proportion of the above-mentioned 0.2 million ha of unimproved grassland is still really biologically interesting. About 50 per cent of it is considered undamaged.

Upland grasslands
Though covering a different ecological range, the great resource of upland grasslands in Britain is unique in western Europe. As there is an overlap with "lowland" habitats, the presence of uplands may partly "mitigate" the loss of lowland habitats. For example Hopkins and Wainwright (1989) estimate the proportion of grasslands with species of calcareous grasslands in upland vegetation as about 0.5 per cent. However, given the large area of uplands relative to calcareous grassland, this may be important. No doubt the position of uplands for acidic grasslands (Nardetalia), at present very rare in the lowlands of the continent, is far more important.

According to Duffey et al. (1974), 36 per cent of the area in agricultural use, 6.4 million ha of Britain, consists of rough grazings. Of this, 1.9 million ha is found in England and Wales, the rest in Scotland. The lowlands in England and Wales only comprise 0.4 million ha, the remaining 1.5 million ha consisting of upland areas.

This upland environment, however, is also threatened. In England and Wales 77 per cent of the enclosed upland grassland received nitrogenous fertiliser with an average of 123 kg N per ha per year. Only 20 per cent received neither fertiliser nor organic manure (Hopkins et al. 1988).

According to Ratcliffe (1984), about 150,000 ha of upland was reclaimed between 1950 and 1980.

Calcareous grasslands
Calcareous grasslands occur on soils, rich in calcium carbonate, like chalk and limestone. Most quantitative information available concerns chalk grassland.

Great Britain, especially England, still has (in international terms), a considerable amount of this habitat. Keymer and Leach (1990) report that 36,682 ha of grassland on chalk remains in England, 23,814 ha in Wiltshire alone. Of the total area, 19,623 ha (53 per cent) is still of high nature conservation interest. However, fragmentation is a problem. Only seven sites are over 250 ha, up to a maximum of 969 ha.

Keymer and Leach (1990) also report that 117,409 ha of chalk downland was present in Dorset in 1793. Already by 1811, only 24.1 per cent remained; in 1934, 6.6 per cent and in 1984, 2.6 per cent. Dorset now holds 3,034 ha chalk grassland, 8 per cent of the total. If the above level of decline is representative, then a total of 1.3 million ha of chalk would have been covered with (no doubt botanically rich) grassland before the Napoleonic war. Of course the loss of downs and heathland must not be so exaggerated that the previous almost complete loss of woodlands is overlooked, but the preservation of a representative area seems justified.

Excellent information about British calcareous grasslands is now available in the proceedings of the 1987 symposium on this theme (Hillier et al. 1990).

No numerical information was available for this survey on conservation measures, but some important instruments do exist, like the designation and management of protected sites such as National Nature Reserves and reserves of the National Trust and County Trusts. In 1984, about 1 per cent of Britain was managed as some form of nature reserve and another 6 per cent had been designated as Sites of Special Scientific Interest (SSSIs) which gives at least a passive protection since the permission of the Nature Conservancy Council is required for certain

Plate 3: Extensive grazing in the Parc Régional du Haut Jura, France [photo: G van Dijk]

Plate 4: Bråsarps Bachar, a species-rich dry grassland reserve in southern Sweden [photo: G van Dijk]

actions or changes of land use. It is intended to give protected area status to some 10 per cent of Britain's land area (Ratcliffe 1984). In 1987 about 8 per cent of Great Britain (*c.* 6,000 sites) was, or would be covered by SSSIs (Wilkinson 1987).

The introduction of management agreements in so-called Environmentally Sensitive Areas (ESAs), which are often SSSIs as well, has been very successful. Although the schemes were introduced from 1987 onwards, 87 per cent of the "targeted land" (119,247 ha) in England was already under agreements by 1988 (MAFF 1989). In the whole of the UK, the ESAs comprised 477,577 ha of farmland in 1988, of which 34 per cent was under agreements (Mathers and Woods 1989).

A high proportion of calcareous grassland is used by the Ministry of Defence. For this survey no information was available on the management and quality of these sites.

Afforestation of rough grasslands may become a problem in future. It is likely that certain vegetation types will be strongly reduced in area.

Ireland (including Northern Ireland)

Until recently Ireland was one of the last large strongholds of semi-natural grasslands. Fertiliser use is among the lowest in Europe, with an average of 47.3 kg N per ha per year in 1982/83 (Logemann 1986). According to Murphy and O'Keeffe (1985) the average use on grassland was 63 kg N per ha per year in 1985. In 11 out of 28 counties it was less than 50 kg and in Leitrim and Longford even as low as 9 and 13 kg. The national average on pastures (48 kg) was lower than for hay (60 kg) and silage (107 kg).

Mulqueen (1988) states that 62 per cent of the Republic of Ireland is grassland. Of a total of 6.9 million ha, 43 per cent is pasture, 19 per cent hay or silage and 14 per cent rough grazing. In the Irish Republic, the landscape is 14 per cent hill (>150 m above sea level) and mountain, 46 per cent dry rolling lowland, 33 per cent wet or mostly wet rolling lowland and drumlins and 7 per cent peatland. Of wet grasslands, only 8.5 per cent (162,000 ha) has been drained, which is considered essential to prevent abandonment. However, according to Baldock (1990) a quarter of the country has been drained in post-war years. This discrepancy may be due to different conceptions of "drainage". Baldock *et al.* (1984) for example state that in post-war years up to 1980, 5 per cent of the farmland had been subject to arterial drainage and 24% to field drainage.

There is detailed information about grassland vegetation in Ireland, including Northern Ireland, from O'Sullivan (1982). The vegetation he lists covers a total of 5.8 million ha, 72.5 per cent of the island. A part of it was still unmanured, including 28,600 ha of calcareous grassland (Brometalia) and roughly 200,000 ha of acid grasslands (Nardetalia), that seem to be of international importance.

Arrhenatherion is restricted mainly to roadsides and cemeteries, and is poorly developed compared with central Europe (average is 23 species, compared to 80 in central Europe). Cynosurion is, or was, very common, with the association *Centaureo-Cynosuretum* encompassing the majority of Irish permanent pastures. It is an association of lightly manured grasslands. Nevertheless at least one of the three subassociations (*galietosum*) seems to be of high conservation interest, as indicated by the presence of species like *Primula veris*, *Pimpinella saxifraga* and *Briza media*. It covered about 670,000 ha, 8.4 per cent of the land area. According to O'Sullivan (pers. comm.) the area of *galietosum* grassland has been greatly reduced in the last twenty years, due to the fact that it occurs on flattish ground (on shallow limestone soils) which is easily cultivated and fertilised. The subassociation *juncetosum* is also of conservation interest. It is often intermingled with (unmanured) Molinietalia communities. Together they covered about 1,565,000 ha, 19.5 per cent of the island. Molinietalia communities comprise Junco conglomeratae-Molinion (Molinion caerulae), Calthion, Juncion acutiflori and Filipendulion. At least a part of this is of high conservation interest. Calthion communities are reported to have suffered severe losses in recent years, due to drainage, fertiliser application and reseeding. Acid grasslands (Nardetalia) occur intermingled with heather-gorse heathland (Calluneto-Ulicetalia). Together they covered about 355,000 ha, 4.4 per cent of the land area. More than half, say 200,000 ha, was considered to be Nardetalia vegetation. O'Sullivan (pers. comm.) states that acid mineral soils in the Irish uplands were planted with conifers in the 1960s and 1970s at which time there was also a move towards the ploughing, levelling, reseeding and fertilizing of heather moorland and acid grasslands. Though substantial areas of these habitats still survive, there is no up to date information on the area they occupy in the landscape. Calcareous grasslands covered about 28,600 ha (0.4 per cent of the island). In the Burren they are widespread on limestone pavements. Elsewhere they are found on eskers, moraines and limestone gravel deposits, where they occur in contact with *Centaureo-Cynosuretum galietosum*. The present area is almost certainly lower due to factors such as scrub encroachment, quarrying and gravel extraction.

In summary some 40 per cent of the grassland in Ireland was still of conservation interest according to botanical criteria, but this is certainly much lower now and the decline is still going on (O'Sullivan pers. comm., Ó'Críodáin pers. comm.). A few sites are now managed as National Nature Reserves. Two pilot areas have recently been set up under the ESA system (Ó'Críodáin pers. comm.).

Germany

Much information has been gathered by individual *Länder* (states), but only a part of it was available for this survey. Some general information is available for the former Federal and Democratic German Republics (GFR and GDR).

According to Grimmett and Jones (1989), 5.5 million ha (22 per cent) of the GFR consists of grassland; in the GDR it is 1.2 million ha (11 per cent). This gives a total of 6.7 million ha (18 per cent) for Germany as a whole. However, according to probably more recent figures (Lee 1989), there is now only 4.5 million ha in the former GFR. Similarly, Lübbe (1988) gives a figure of 4.6 million ha for the GFR, 38 per cent of the agricultural land. He estimates that only 3 per cent of the permanent grassland consists of extensive forms of grassland, like scattered meadows, alpine pastures and rough pastures. Lübbe (1988) also estimates that up to 1 million ha of permanent grassland will become redundant. Marginal grassland locations will be given up first and it is proposed to redevelop these areas into natural biotopes.

According to H. Schulzke (pers comm.) the proportion of grasslands of the farmland in Nordrhein-Westfalen has declined by about 33 per cent in the period 1965-1986, whereas the total agricultural area declined by 21 per cent. The average use of nitrogenous fertiliser in agriculture in the GFR was 108.5 kg per ha per year in 1983 (Logemann 1986).

Korneck and Sukopp (1988) have listed the numbers of endangered species in the former GFR and the proportions per habitat type. Dry and wet grasslands turned out to be amongst the most endangered habitats. The number of endangered species of higher plants and ferns was an astonishing 727. The number of species of dry and semi-dry grasslands (Festuco-Brometea) was as high as 588, but 201 of them (34 per cent) were endangered. For wet grasslands (Molinietalia) 297 species were mentioned, of which 96 (32 per cent) were endangered. In well drained neutral grasslands (Arrhenatheretalia), 240 species occur, of which, strangely enough, only 25 (10 per cent) were reported to be endangered, whereas this habitat type is declining very quickly.

A provisional totalling of the data for calcareous grasslands gives an area of almost 40,000 ha (including at least 10,000 ha abandoned land) or more for the former GFR, excluding Rheinland-Pfalz, Saarland and Hessen, which is no doubt of great international importance.

Nature conservation is a responsibility of the *Länder*, and today considerable efforts are made to preserve the remaining semi-natural grasslands. Over 100,000 ha of grassland is now under agreements (Ziesemer, pers. comm.), either for botanical or faunistic reasons. This is 2.2 per cent of the total grassland area in 1990.

More information will now be given about various parts of Germany.

Schleswig-Holstein

With 482,000 ha of grassland the proportion of grassland is 44 per cent of the agriculturally used area. This area has hardly changed since 1950 (Ziesemer 1989). About 7,500-11,000 ha, 2 per cent of the grassland, is still of botanical interest (Ziesemer, pers. comm.).

There is a "red list" of threatened plant communities (Dierssen *et al.* 1988) which gives distribution maps of the communities and descriptions of their status but without the area of each. It is stated that 70 per cent of the plant communities are threatened and 44 per cent of the species. The main causes of loss are eutrophication of the landscape and disappearance of traditional farming.

Most vegetation types of conservation interest now cover only small areas. However, wet grasslands are still relatively well represented, in spite of a great decline.

Management agreements are possible on no less than 252,000 ha. However, only 16,500 ha of this (consisting of 76 per cent wet and 24 per cent dry grassland) is of potential botanical importance.

Nordrhein-Westfalen

Detailed information is available for about one third of this *Land* (Landesanstalt für Ökologie, Landschaftsentwicklung und Forstplanung 1988), covering the southern rather mountainous region of Mittelgebirge.

About 9,000 ha is considered to be of conservation interest, of which 1,000 ha is abandoned grassland. Calcareous grasslands cover 926 ha, mostly located in the Eifel, of which a great part is protected. Heavy metal grasslands cover 113 ha, acid grasslands (Nardetalia) 38 ha, neutral hayfields (*Fettwiese*) 980 ha, neutral pasture (*Fettweide*) about 3,000 ha. Though the calcareous and heavy metal grasslands are of more than regional importance, semi-natural grasslands as a whole now occupy only a fraction of their former area.

H. Schulzke (pers. comm.) reports a great decline of species-rich grasslands. Nardo-Galion (acid grasslands), Arrhenatherion (neutral), Trisetion (neutral, higher altitudes) and Calthion (wet) have greatly diminished. East of the Rhine the status of Arrhenatherion is even worse than in the Eifel, with certainly less than 10 per cent left. In the flat and hilly parts of the *Land* less than 1 per cent of Calthion is left and Nardo-Galion and Festuco-Sedetalia have practically disappeared. In the floodplains of the Rhine, *Medicagini-Avenetum* (Mesobromion, so related to calcareous grasslands) was rather common about 1960, but has now almost disappeared.

Today, great efforts (e.g. management agreements) are being made for protection in the framework of the *Mittelgebirgsprogramm, Feuchtwiesenprogramm, Natur 2000 in NRW* and *Programm zur Wiedereinfuhrung historischer Landnutzungsformen*.

Lower Saxony (Niedersachsen)
The area and distribution for various vegetation types or combinations of them is given in the *Naturschutzatlas Niedersachsen* (Von Drachenfels *et al.* 1984). Calcareous grassland covered 767 ha, scattered over 145 sites. This fragmentation shows that much more must have been present in the past, though only in the southeast corner of the *Land*. Recent investigations show that about half of this area has disappeared in recent years (Von Drachenfels 1988).

Acid semi-natural grasslands, i.e. Corynephorion, Thero-Aïrion, Armerion (elongatae?) and Violo-Nardion, covered only 1,048 ha, scattered over 110 sites. This rarity is no doubt due to afforestation of heathland and agricultural intensification.

Heavy metal grasslands covered 38 ha, but it is not likely they ever were common. Neutral hayfields over 300 m above sea level (Polygono-Trisetion and transitions to Violo-Nardion) covered 296 ha. Due to topography they only occur in the southeast. Arrhenatherion was not mentioned, but species-rich forms are no doubt rare.

Wet or moist grasslands covered a much larger area of 20,860 ha, plus a part of the 25,595 ha *Talniederung* (valleys), in total 0.5-1 per cent of the land area. This somewhat better situation is due to the character of Niedersachsen. However, recent investigations (Von Drachenfels 1988) show that the greater part of these areas are now poor in species.

The decline of grassland seems little different from other types of vegetation. In 4,000 ha of the Dümmer lakeside, for example, Ganzert and Pfadenhauer (1989) found a decline of all species-rich types from over 90 per cent coverage in 1947 to 2-3.5 per cent in 1987.

Baden-Württemberg
Information about the results of the inventory of this *Land* was obtained from B. Schall (pers. comm.). This concerns mainly calcareous grasslands and acid grasslands.

Ungrazed calcareous grassland covers about 4,000 ha, but a great part has been abandoned. Grazed calcareous grassland covers some 18,500 ha, but 60-70 per cent has been abandoned. Detailed examples are given of the decline in certain areas, varying from around 50 to 90 per cent, mainly due to succession.

Acid grasslands (Nardetalia, mainly Violion) cover about 11,500 ha. Arrhenateretalia grasslands rich in *Salvia pratensis* cover some 1,000 ha.

O. Wilmanns (pers. comm.) reported that a large part of the remaining calcareous grasslands are protected and that more intensively used hayfields of Arrhenatherion and Trisetion are more threatened. In the Black Forest the use of liquid manure is said to be an important cause of decline.

An important site for calcareous grasslands is the *Kaiserstuhl*, where about 150 ha has been protected. There has been a considerable increase of the vineyard area, mainly at the cost of calcareous grassland and (in particular) neutral hayfields (Arrhenatherion) (Wilmanns 1989).

A very important series of calcareous grasslands is still present in the Suabian Jura (*Schwäbische Albprogramm*). Briemle (1988) reports that 30,000 ha remains. There are still some 100,000 sheep in this area. In a part of this area 3,580 ha (48%) disappeared between 1900 and 1980, 47 per cent of the loss was due to scrub encroachment, 26 per cent to afforestation, 17 per cent to intensification and 10 per cent to other land use. In the framework of the *Albprogramm* the government of Baden-Württemberg has been trying to preserve this landscape for more than 25 years by supporting sheep rearing.

Bavaria (Bayern)
The first survey was made in 1974-1977. It is now being repeated, but not yet completed. Data from the 1970s were provided by E. Wenisch (pers. comm.).

Calcareous grasslands cover 15,776 ha. Acid grasslands (vegetation type unknown) cover 4,409 ha. Complexes of vegetation with wet grasslands as important components (*Wiesentäler, Flachmoore und Streuwiesen, Nasswiesen*) cover almost 45,000 ha.

J. Heringer (pers. comm.) mentions 17,000 ha of "old pasture", extensively grazed areas, mainly calcareous grasslands and acid grasslands, which is mainly a residue of common pasture.

Southern Germany is likely to be a part of the former centre of well-developed Arrhenatherion vegetation. However, no information is available about the status of these vegetation types. Information from Nordrhein-Westfalen (H. Schulzke, pers. comm.), Baden-Wurttemberg (O. Wilmanns, pers. comm.) and Austria (personal observation) suggests that important losses have taken place here, too.

Bavaria has protection programmes comparable with those of other *Länder*. About 40 per cent of the entire land has been designated as ESA (Baldock, this volume).

In Bavaria grazing of forests is considered harmful and efforts are being made to separate forests and pastures. Since 1960 at least 14,000 ha of forest has been freed from grazing rights (Prieß and Haxel 1982). Elsewhere, however, integrated grazing is considered interesting from a nature conservation viewpoint. Further information on this subject would be useful.

Former German Democratic Republic
Klafs *et al.* (1979) describe the situation in Mecklenburg, which had become totally transformed from the former grassland situation. Rutschke *et al.* (1983) describe a similar situation for Brandenburg. Formerly extensive Molinion vegetation had disappeared during the last 20 years.

Recent information from the university of Halle (S. Klotz, pers. comm.) confirms this picture. In the GDR economy, cereals were very important and grassland had mainly been reseeded with highly productive varieties. The extensive peat grasslands in the north have been drained and destroyed for a long time. Riverine grasslands in the south have been turned into arable land or intensively used grassland. Somewhat better is the situation of calcareous grasslands. These could not easily be intensified by collective farms and were extensively used by small private holdings. However, many of them were abandoned and subject to scrub encroachment. In the 1980s collective farms enlarged sheep stocks, which was favourable for the management of dry grasslands. However, it is very doubtful if this situation will be maintained.

Böttcher and Schlüter (1989) describe vegetation changes in a river flood plain in Saxony (Sachsen). At the end of the 1960s 17 types of grassland vegetation could be distinguished. Only four now remain, often as small, fragmentary remnants. One third was transformed to arable land. Of the remaining grassland 12 per cent was considered to be "traditional" types.

France
France has about 12 million ha of grassland (Lee 1989), 21 per cent of the country. The average use of nitrogenous fertiliser in agricultural use in 1982/83 was 69.6 kg N per ha per year (Logemann 1986). According to Delpech (pers. comm.) the great majority of the grasslands receive less than 50 kg N per ha per year and have a spontaneous, stabilised vegetation, and are considered semi-natural by him.

France is known for some important strongholds of semi-natural grasslands, like Vosges, Jura, Causses, Normandy, Auvergne, Aquitane, parts of Languedoc and Provence, Quercy, Alps, Pyrenees and some river valleys. No doubt several areas are of great international importance. However, exact data are very scarce and negative changes take place very quickly. Though little is known about the exact status of the areas involved, useful information about dry and semi-dry grasslands is available in the proceedings of the international symposium on calcareous grasslands in 1982 (Géhu 1984). Similar books exist on flood plain grasslands (Géhu 1978a) and therophyte grassland (Géhu 1978b).

Very interesting information about several regions, like Burgundy (Bourgogne), Champagne-Ardennes and Franche-Comté, including a great number of river valleys was received from J. M. Royer (pers. comm.). This information shows that France still supports grassland of exceptional quality in western Europe.

Although no good survey of France exists at this moment, a description of 14,000 areas of conservation interest is under preparation and is planned to be available in two years (Sécretariat de la Faune et de la Flore, Muséum National d'Histoire Naturelle; H. Maurin pers. comm.)

The Jura
Wolkinger and Planck (1981) mentioned the Jura as one of the most important areas in France for dry grasslands. This is confirmed by information obtained from Royer (pers. comm.), concerning some tens of thousands of hectares of species rich grasslands of different alliances.

According to Pottier-Alapetite (1943) in 1930 39 per cent of valleys (*plaine*) in the Doubs region, excluding forests, was "natural grassland" (*prairies naturelles*). On the plateaux (*moyens plateaux*) this was 43 per cent. It is remarkable that there has been an upward trend, for in 1882 these figures were 21 per cent and 7 per cent. In 1930 "artificial grasslands" covered respectively 10 per cent and 7 per cent of the areas.

The natural grasslands mainly comprised *Arrhenatheretum* in the valleys and mainly *Mesobrometum* on the plateaux. The figures suggest that over 200,000 ha were covered with these associations. 30,000 ha consisted of very poor communal grazings, including some xerophilous grasslands (probably *Xerobrometum*). According to Willems (pers. comm.) by 1970 much of this was still intact, but by 1976 a very great part of the calcareous grasslands had disappeared. There were all kinds of causes: agricultural intensification, afforestation, building of houses and camping sites. Therefore it is most urgent to investigate the present situation and start conservation measures in the immediate future.

The southern part of the Jura has higher altitudes than the rest, and has been designated as a regional park. The great majority of the grassland is still of a very high quality. In spite of the altitude, the vegetation is rich in plants of calcareous grasslands. According to J.M. Royer (pers. comm.) the grasslands between 900 and 1,200 m above sea level belong to Mesobromion, Cynosurion and Arrhenatherion, with transitions to Seslerion and Polygono-Trisetion. The map of the regional park mentions an area of 62,088 ha, of which 36,631 ha is covered by forest. The majority of the rest, 25,457 ha, is covered by pastures and meadows. A rough estimation is that three quarters of this or more is still rich in species. Outside the park we find similar situations (Doubs, Ain; M. Forestier, pers. comm.). This makes it one of the largest more or less contiguous areas of semi-natural grassland in the continental part of western Europe. So far the regional park can give no effective protection to the grasslands (M. Forestier pers. comm.). Because of European regulations the farmers cannot raise productivity further. In fact there is only one choice for further development: quality. At present, the production of regional cheese brands (Comté, Bleu, Gex) means the use of fertiliser is not forbidden, though other restrictions exist. Opinions differ about the necessity of having a semi-natural vegetation.

Much of the grasslands in the park (perhaps one half) have been abandoned as meadows but most of it is still grazed by cattle from Switzerland. If this suddenly ceased, for example because intensification elsewhere made these grazings unnecessary, succession to forest would take place quickly. However, at this moment not more than an estimated 10 per cent has really been abandoned.

Though the present situation in the grazed areas (probably not fertilized) is very favourable, it is unstable as well. In a recent newspaper article it was stated that 32 per cent of the cattle farms in France had disappeared during the last six years. So, immediate action is very necessary. The introduction of management agreements, together with measures for stimulating ecological farming, seem appropriate. At least the most vulnerable parts, for instance damp valleys with rare species, merit protection as nature reserves.

The Loire valley
This valley is famous because of its relatively natural character, unique in western Europe. It is also one of the most important areas in Europe for *Fritillaria meleagris* which occurs in the association *Frittillario-Oenathetum*, in the Arrhenatherion (Corporaal 1990). The number of plants was estimated to be about four billion and more than sixty times this in the seedbank. However, at the same time he estimates the number of individuals to be only 10 per cent of the level of 20-30 years ago, which illustrates the urgency of conservation measures.

The upper Loire valley is also important. J.M. Royer (pers. comm.) specially mentions the part between Digoin and Charité sur Loire, where grasslands are increasingly abandoned.

Other river valleys
Though the present information is incomplete, some valleys deserve special mention: Vallée de l'Oise (Géhu 1978a), Bassin de la Voire (vallées de la Voire, de l'Héronne, de la Laines) (Royer and Didier 1982), Vallée de la Saône (J.M. Royer, pers. comm.).

The Saône valley between Gray and Mâcon comprises over 15,000 ha and c. 95 per cent is still semi-natural and of a high botanical diversity (J.M. Royer, pers. comm.). Protection measures are very urgent here.

According to J.M. Royer (pers. comm.) similar sites are found in the valleys of the Marne, Seine, Aube, Aisne, Meuse and Chiers: large areas (>10,000 ha) with a great majority (95 per cent) of species-rich grasslands, but rapidly declining due to other land use.

This also applies to a number of little valleys in the calcareous area of Champagne-Ardennes, with over 5,000 ha of well developed grasslands (J.M. Royer, pers. comm.)

Vosges/Alsace
Along the western border of the Alsace there is a series of pastures (*chaumes*) above the forest zone. They cover about 5,000 ha, the majority being secondary, but 270 ha is considered natural (J.R. Zimmermann, pers. comm.).

In the last century the area was about 6,500 ha and some 5,000 cattle grazed there. Now there are about 2,800 cows and 1,200 sheep on 5,000 ha. The decrease was partly due to the first world war, when the battle front passed the area (J. R. Zimmermann, pers. comm.)

Though the soil is not calcareous, the meadows are very important botanically. High numbers of *Arnica montana* give a beautiful aspect to certain parts. Tourism (*fermes-auberges*) helps the farms to survive so far, so that abandonment is no immediate threat. However, intensification, though still on a limited area, has appeared here.

Other regions with semi-natural grasslands
Wolkinger and Planck (1981) give information about the most important regions for calcareous grasslands: Jura, pre-Alps, Quercy and the Causses. A map shows the distribution of numerous other sites and areas in France. They report that in the Champagne region, 99 per cent of the dry grasslands disappeared between 1950 and 1980. Moreover, dry grasslands are generally shrinking very rapidly, because of land consolidation, ploughing, etc. The only protected area was the Cévennes, where dry grasslands are part of a national park.

To the areas with important concentrations of semi-natural grasslands, the following regions must be added (R. Delpech, pers. comm.): Aquitane, parts of Languedoc and Provence and the middle and lower Rhône valley.

In northwest France there are several areas of calcareous grasslands. Scott (1971) and Van Speijbroek (1984) describe the Somme valley, with Mesobromion vegetations probably older than our post-glacial period. Géhu *et al.* (1984) describe the Boulonnais, where calcareous grasslands showed a marked decline, due to both abandonment and intensification.

Barbero *et al.* (1984) describe the difficult situation of dry grasslands in the southeast. Grazing and controlled fires are considered necessary for this vegetation. Abandonment is a problem. Seven alliances in need of protection are mentioned.

Pautou and Girel (1984) describe the decline of

Mesobrometum and *Xerobrometum* on permeable soils in the Rhône valley, threatened with imminent disappearance.

Duvigneaud (1984) describes the almost complete disappearance of a special subassociation of Mesobromion from the Moselle valley, due to fertilizing, land consolidation, conversion to maize crops and poplar plantations, gravel extraction and other causes.

As was reported earlier for Germany, *Arrhenatheretum* is quickly declining there. There are indications that the situation in France is better. Arrhenatherion is well represented in the areas mentioned by J.M. Royer listed above. According to P.H. Veen (pers. comm.) it also still occurs commonly in at least a part of the French Alps.

Grassland conservation in France
So far, very little has been done for the preservation of semi-natural grasslands in France. As far as is known, only four sites are currently protected (Delpech, pers. comm.). During the symposium on calcareous grasslands (Géhu 1984) a proposal was developed to establish a biogenetic network of calcareous grasslands throughout Europe (Géhu 1984), in accordance with the recommendations of Wolkinger and Plank (1981) but up to now no international activity has been generated by this proposal (Willems 1990), which is especially regrettable concerning France, where the symposium was held.

Recently a start has been made with the introduction of management agreements. Priority is given to wetlands, to the struggle against forest fires and to aid for pastoralism. In the Marais de Poitevin agreements are now possible on 2,000 ha, financed by the EC, the Ministry of the Environment, the Ministry of Agriculture and regional authorities, with further support from Worldwide Fund for Nature (WWF) and the Ligue pour la Protection des Oiseaux. Another opportunity is the EC support for farmers in less favoured regions. However, an important area like the regional park in the Jura is not yet involved. Moreover the amount of grant per hectare is low and no agreements are made as to the type of management.

Switzerland
At the moment the only information is about calcareous grasslands *(Trockenrasen)*. Hedinger (1984) states that about 90 per cent of this vegetation had already disappeared by 1984: there may even be less now, for developments can go very quickly, as is illustrated by data from the Jura canton. In 1981 only 32 per cent of the sites, known in 1975, remained; 59 per cent of the loss was ascribed to fertiliser use. In the remaining sites many rare species had disappeared (Klein and Keller 1983). According to the same source, over 90 per cent of the sites known in 1968-1971 in the Soloturn canton had lost their value.

Intensification of agriculture and abandonment were important causes of the decline.

Broggi (1989) lists the area of calcareous grasslands in the cantons of Solothurn, Aargau and Bern: repectively 150, 682, and 111 ha, which means 1.5, 1.0 and 0.3 per cent respectively of the farmland. Given the bad situation of Mesobromion grasslands, the situation of *Arrhenatheretum* grasslands would hardly be any better.

As we saw, the difference between Switzerland and the adjoining French Jura is striking.

In Switzerland efforts are made to save dry grasslands by means of management agreements, for example in the Solothurn Jura. Here management agreements for three different types of hay meadows are possible and a further one for pastures. Restrictions are imposed on the use of fertilisers and manuring (none on Mesobromion) and the date of mowing.

Austria
Wolkinger and Plank (1981) describe a great decline of dry grasslands in the Neusiedler See region, near the Hungarian border, due to conversion to vineyards (almost 4,000 ha).

An excellent survey of Austrian dry grasslands and their rich flora and fauna is given by Holzner *et al.* (1986).

Schratt (pers. comm.) reports that hay meadows have become very rare in regions suitable for arable land. Calthion vegetation is even more threatened than Mesobromion and Nardion vegetation. Apart from conversion to arable, meadows are mainly threatened by eutrophication (fertilisers) and abandonment.

Personal observations in the Tirol showed that species-rich grasslands, including *Arrhenatheretum*, seem to have become great rarities. Intensification on flatter parts and afforestation on steep slopes were observed as causes.

Grabherr (pers. comm.) reports that a large part of the remaining Pannonic dry grasslands are now protected (see Kollar, this volume). Saline meadows in the same region have also declined considerably but some have been protected. Well developed Mesobromion grasslands are still present in the Wienerwald, Kärnten, Steiermark and Tirol, but all are very threatened. In addition high altitude grasslands are now threatened, both by intensification and abandonment.

In the big Alpine valleys, wet meadows *(Streuwiesen;* mown in autumn) were formerly important. In the Rhine valley in Vorarlberg, formerly the most important area for this vegetation, two-thirds have disappeared, but 800 ha of these meadows are likely to be protected (Grabherr pers. comm.)

The *Länder* are responsible for nature conservation. Many efforts are made to safeguard species-rich grasslands. The current GATT negotiations and Austria's position towards the EC will be decisive for the future (Grabherr, pers. comm.)

Belgium

Belgium holds 632,000 ha of grassland (Lee, 1990). Within a few years a survey is expected for Flanders (Kuijken, pers. comm.), where the area of semi-natural grasslands is rapidly declining (Kuijken, 1988).

Wolkinger and Plank (1981) report an 80 per cent decline of dry grasslands in Belgium over 150 years. Noirfalise and Dethioux (1982) mention a total of 1,715 ha of protected sites, a quarter to one-third of this is actually occupied by dry grassland. According to Willems (pers. comm.) scrub encroachment is a problem in Belgian calcareous grasslands.

Denmark

Denmark holds 214,000 ha of grassland (Lee 1990). According to Wolkinger and Plank (1981) dry grasslands have shrunk greatly since the 1950s. A large part of that which remains is protected, such as in the north of Jutland. No other information is available.

Denmark has introduced "Article 19 schemes" for ESAs (Baldock, this volume).

Sweden

Some statistical information has been published by Naturvårdsverket (1987). The total area of farmland, belonging to farms with more than 2 ha of *åker*, the most intensively used land, was 3,247,000 ha in 1986. 2,908,000 ha of this was *åkermark*, including 859,000 ha grassland (*vall*). Outside this *åker* category another 340,000 ha was pasture (*betesmark*). In 1976 the total area of *betesmark* in holdings with any *åker* was about 730,000 ha, of which about 250,000 ha was "cultivated pasture" and about 480,000 ha "natural pasture".

It is this last category which is of most interest here. Out of the 480,000 ha, 53 per cent belonged to holdings with less than 2 ha of *åker*, which seems to illustrate the danger of a sudden disappearance.

Sweden is famous for its park-like meadows, known as *ängar* (singular: *äng*). The grazed variant is called *hage*. In many cases they have scattered trees or shrubs, plots of woodland and glades, which gives them a very special character. The remaining areas are often of great botanical importance, so that it may be assumed that the strong decline in the past meant an important ecological loss.

The average use of fertiliser (all crops) is about 90 kg N per ha per year. However, *ängar* and *hagar* (10-20%) are hardly fertilized. Currently, an inventory of *ängar* and *hagar* is being carried out. (K. Ekeland, pers. comm.)

According to Gustawsson (1976) the area of *åker* in Sweden increased from 850,000 ha around 1800 to 3.5 million ha around 1900, mainly at the cost of *ängar*. Naturvårdsverket (1987) estimates the area of "natural grassland", excluding rough grazings such as *alvar*, heathland and forest at about 2 million ha around 1850 and 0.5 million ha in 1950, whereas the area of *åker* rose from 2 million ha in 1850 to almost 4 million ha in the first half of this century.

Important areas for dry grasslands are the islands of Öland and Gotland and the small islands of Store and Lille Karlsö. An exceptional case is the huge dry grassland area Stora Alvaret (*c.* 25,000 ha) in the island of Öland. An important part of the vegetation belongs to the Brometalia order. Up to 40 plant species per m^2 and up to 30 per 0.25 m^2 are found, with almost equally high figures for mosses and lichens, and a total of 388 plant species on the *alvar* (Van der Maarel 1988).

G. Regnéll (pers. comm.) reports that formerly fertilization, ploughing and draining were the greatest threats for semi-natural grasslands, whereas pastures with a low productivity were mainly threatened by afforestation and natural succession. Today, there is a general threat of afforestation, natural succession, other land use (construction etc.) and nitrogen-rich precipitation.

K. Ekeland (Naturvårdsverket: pers. comm.) states that (biologically) valuable grasslands are threatened both by intensive and "extensive" use. In general, the greatest threat is abandonment.

Today many small areas and certain types of open grasslands are protected as nature reserves. However, no data are available about what proportion this is of the total. According to K. Ekeland (pers. comm.) the majority of *ängar* and *hagar* is in private ownership. Since 1986, a scheme for management agreements, "NOLA", has been in existence (40 m crowns per year) and in 1990 a similar scheme ("Landskapvård"; 100 m crowns) has been introduced (K. Ekeland, pers. comm.).

Norway

Losvik (1988) has published an interesting study of meadows in Hordaland, western Norway. All the communities she describes are assigned to the Arrhenatheretalia, with occasional transitions to Nardetalia and Molinietalia. Eight units are described, four of the Atlantic alliance Cardaminion pratensis, one of the alliance Arrhenatherion (few localities) and three provisional groups.

No meadows in the study received more than 120 kg N per ha per year. The lowest quantity was applied on steep slopes. Well fertilized meadows receive a lot of manure in addition to at least 120 kg N per ha (Losvik, pers. comm.).

Losvik (pers. comm.) estimates that 10-20 per cent of the grasslands in the study area are still botanically interesting, whereas this may have been 90 per cent in 1950. No doubt fertilizing has played an important role in the decline but Losvik (1988) states that abandonment is probably the most common reason for changes in meadows at present. In developing areas or in remote districts whole farms are left fallow and within managed farms steep or shallow land is often abandoned to increase efficiency.

Italy

Little information was found for Italy in the time available. According to Grimmett and Jones (1989) there is 5,000,140 ha of permanent pasture in Italy, which differs little from the 4,944,000 ha (28.3 per cent of farmland) mentioned by Lee (1990). Logemann (1986) mentions an average use of 55.9 kg N per ha per year in 1981-1982 in the total agricultural area.

Wolkinger and Plank (1981) state that sub-Mediterranean grasslands (Brometalia erecti) originate from the calcareous mountains of the northern Mediterranean, in particular northern Italy, southern France and Spain. They refer to very good information about areas of conservation interest in Italy. The Valle d'Aosta and Trentino-Alto-Adige region are especially mentioned as regions with important dry grasslands, whereas the map shows seven larger areas in the Alpine region and several areas in the Apennines, mainly in the upper half of the peninsula. The calcareous grasslands of the central and southern Apennines are assigned to the endemic alliance Crepido lacerae-Phleion ambigui (Blondi and Blasi 1984).

A large proportion of the steppe areas of southern Italy (Apulia, Sardinia, Sicily) have been replaced by cultivated fields during the last 100 years and especially during the last 30 years (Grimmett and Jones 1989). The area of arable land in these regions increased from 283,473 ha in 1958 to 1,053,873 ha in 1983. According to Petretti (this volume) 200,000 of steppe habitats are still left in these regions and Latium.

Spain

According to Lee (1990), Spain holds 6,645,000 ha of permanent grassland, 24.4 per cent of the country's farmland. According to Grimmett and Jones (1989) this farmland covers 53 per cent of the country. According to Baldock and Long (1987) "natural meadows" cover 1,452,300 ha and pastures 5,193000 ha, making 6,645,300 ha altogether. Moreover, they mention 3,660,500 ha of common land grazing and 420,900 ha of *Esparto* grassland.

According to Baldock and Long (1987) the average use of nitrogen fertiliser was said to be 53 kg nitrogen per ha per year according to spanish sources, whereas the EC Commission mentioned 32 kg per ha per year. They suppose the first figure applies to arable land only, which would mean that fertiliser use on grasslands was very low if 32 kg was the average.

Some three million hectares have been reafforested with *Pinus* and *Eucalyptus* and large-scale replacement of pasture land by *Eucalyptus* plantations are expected in the near future (Grimmett and Jones 1989).

Wolkinger and Plank (1981) mention large areas of calcareous grasslands in the Pyrenees, Asturia (Cantabrian mountains) and Javalambe. De Juana *et al.* (1988) describe steppe habitats in Spain and current threats, but no numerical data are given.

Ruiz and Ruiz (1986) state that *transhumance* practices, the complementary exploitation of highlands and lowlands, for which a network of livestock routes was used, have led to one of the most complex and interesting ecosystems of the Mediterranean. According to Bignal (this volume) this system is essential for many birds such as bustards, vultures and eagles. Though migratory movements still continue, they have been drastically reduced.

Very interesting from an ecological viewpoint are also the *dehesas* or tree pastures. Ruiz Perez (1986) estimated that five years ago only 2.3 million ha of *dehesas* were left in Spain and Portugal combined. Though, according to Baldock (1990), the area of this habitat (in Spain) has shrunk rapidly from 4.6 to 3.4 million ha over the period 1970-1984.

In their studies on the expected consequences of Spain's EC membership, Ruiz Perez and Groome (1986) predicted intensification of agriculture in the most fertile regions and marginalization and abandonment in many less productive areas. Baldock and Long (1987) reached the same conclusions in their study. They concluded that payments to farmers to continue traditional forms of production which are no longer viable are undoubtedlty desirable from an environmental viewpoint. They recommended the raising of the EC contribution to "Article 19 schemes" for Environmentally Sensitive areas to a higher level.

Less Favoured Areas cover 17 million ha according to Baldock and Long (1987), over 60 per cent of the agricultural area. They also mention the potential EC contribution of 80 per cent of the costs of afforestation in Less Favoured Areas at that time. They suppose more marginal land is likely to move out of agriculture and become available for afforestation.

The potentially most damaging developments identified by Baldock and Long (1987) are: intensification of agriculture, the abandonment of traditional systems and inappropriate forms of forestry.

Ruiz perez and Groome (1986) foresee similar consequences of Spain's EC membership: "a dichotomy between the intensification of agricultural cultivation in the most fertile agrarian regions... and the large-scale marginalization and abandonment of many less productive areas". According to these authors conversion to forest plantations of fast growing species is one of the most likely and at the same time most ecologically and socially dangerous outcomes for many of these areas.

Portugal

According to Lee (1990) Portugal holds 761,000 ha of permanent grassland, 16.8 per cent of the country's farmland. Grimmett and Jones (1989) state that afforestation of poor farmland and "wasteland" is having a major impact on the landscape. A five-year programme (started in 1982) included the planting of 1,360,000 ha with *Pinus* (two-thirds of the area) and *Eucalyptus*.

According to Gaspar *et al.* (1988) Portuguese agriculture is changing very slowly and remains at a low degree of intensity. "Natural pastures" are of poor quality and can support only 0.2-2 sheep/ha without liming and fertiliser application. In the south woodland grazing is very important.

It may be supposed that the low intensity of agriculture has favoured botanically interesting areas, but no data were available for this survey.

Baldock and Long (1987) expect similar effects of agricultural modernization to those in equivalent situations in Spain: the twin pressures to intensify production on the better land and to cut back traditional and marginal systems and also similar pressures to accelerate afforestation. They quote Portuguese sources that suggest that around two million ha of marginal farmland should be transferred to forest.

Greece

According to Lee (1990) Greece has 1,789,000 ha of permanent grassland, 31.2 per cent of the country's farmland. Though no description of grassland vegetation was available for this survey, the very rich flora of Greece, with 750 endemic species (Committee of threatened plants 1982), gives a good indication that Greece should not be neglected. In 1981-1982 the use of nitrogenous fertliser, at 36.5 kg N per ha per year (Logemann 1986), was the lowest in the then EC territory.

Eastern Europe

Information from eastern Europe was recently collected by IUCN (1990), but mainly for ornithological purposes. For some countries additional data have been collected and included here.

A general impression is that vegetations have suffered most in countries with collective farming. This is derived from data about the former GDR (DDR), which were presented above. Two countries (Poland and Yugoslavia) differ greatly from the others by the fact that their agriculture is predominantly private. At least in Poland it is known that semi-natural grasslands still cover large areas, mainly in the lower parts of river valleys. Though parts of these areas may have been in common use until recently, this use is different from the rational use in collective farms.

Poland

According to IUCN (1990), Poland has 4,040,000 ha of grassland (90 per cent lowland grassland) which is 13 per cent of the country and 22 per cent of the farmed area. Three-quarters of all cultivated land is private. According to Grimmett and Jones (1989), 2.3 million ha have been changed by drainage and other projects. The rest was classified by IUCN as damp and wet grasslands but in fact dry grasslands are a common component of flood plain systems (personal observation).

Personal impressions (Van Dijk 1990) were that outside of floodplains, agriculture is usually intensive enough to have changed vegetation drastically, but that river valleys still hold extensive areas of semi-natural grassland. Along larger rivers these are grasslands on mineral soils (e.g. Armerion elongatae on drier parts, Molinietalia on the lower parts): along smaller rivers (e.g. Narew and Biebrza) they occur on peaty soils. They are of a different character and still retain a very great botanical importance (e.g. partly Caricion davallianae, P.H. Veen pers. comm.). However, large areas of such grasslands have recently been abandoned and the remaining part must be regarded as extremely threatened. The combination of intensification of the better land and abandonment of the rest is not restricted to western Europe, but also occurs here. The recent great agricultural problems may even hasten this process.

Wolkinger and Plank (1981) show a large area in the south east of Poland where calcareous grasslands occur. In fact calcareous grasslands have become extremely rare here, probably mainly due to conversion to arable land. The few nature reserves are not mown or grazed and are likely to lose their wealth of grassland species if succession continues.

Recently a very detailed description of Molinio-Arrhenatheretea grasslands (and two other classes) in the Lublin area, about 10 per cent of Poland, was published (Fijałkowski and Chojnacka-Fijałkowska 1990). The total area of grassland described was 222,000 ha. The authors state that almost all types, except for *Molinietum medioeuropaeum* (*c.* 70,000 ha) is of very high economic value. This latter category also comprises 48,000 ha of *Arrhenatheretum medioeuropaeum*. At this moment it cannot be said what proportion of these vegetation types is still in a good botanical condition and what parts have been changed by intensification or abandonment. Though the general picture, as shown before, may apply to this region, too, the *aufnahmen* (*relevées*) show that, when they were made, many interesting species were still present, especially in *Molinietum*. As for dry grasslands in river valleys, these are at least locally very well developed along the Bug near the USSR border.

Plans are being developed for a national survey of river valley grasslands in Poland, which would be the first step towards a protection policy of semi-natural grasslands, which, in contrast to natural vegetation, have, been relatively neglected so far.

Polish grasslands are also important for birds that have become rare in the west, like white stork *Ciconia ciconia* (some 30,000 pairs according to Grimmett and Jones 1989), whinchat *Saxicola rubetra*, and cranes *Grus grus*.

It seems that Poland has a very important position in Europe, mainly because of its river valleys. However, the remaining semi-natural grasslands are seriously threatened by abandonment, intensification and some conversion to arable, and the lack of management in protected areas.

Yugoslavia

According to Grimmett and Jones (1989) Yugoslavia has 6.4 million ha (25 per cent of the country) of meadow and pasture, including numerous unimproved flood meadows, mainly in the northern half of the country. Their inventory includes a number of very important areas of flood-meadows and riverine forest along the rivers Sava, Drava and Danube in Croatia and Vojvodina, but many of these sites are unprotected.

According to Stanley (1989) in Yugoslavia, 85 per cent of the agricultural land is private. This, in combination with a low intensity of use, is probably an important reason for the presence of many ecologically rich grasslands.

In the list of important bird areas (Grimmett and Jones 1989) some very large river valleys occur, covering hundreds of thousands of hectares.

Schneider-Jacoby (1990) states that the river Sava is still in a good condition over almost its total length of 900 km. Because of water fluctuations of nine metres, up to 790,000 ha were inundated up to 1970. In the framework of a big amelioration project, "Sava 2000", 90 per cent of this has been drained. Nevertheless, a very important area has been protected recently: the Nature Park Lonjsko Polje, covering some 50,000 ha (L. Ilijanic, pers. comm.). According to Schneider-Jacoby the area includes 36,000 ha of alluvial forest and 12,000 ha of meadows and pastures, used in a traditional manner. One of the threats to the area is pollution from the industrial towns of Zagreb, Sisak and Kutina.

According to Ilijanic (1973), the meadows of the lowlands in the Sava valley with fluctuating humidity (*wechselfeucht*) belong to the alliances Molinion in the north, Deschampsion cespitosae in the middle part and Trifolion pallidi in the southeastern part. Other meadows belong to Arrhenatherion (Ilijanic 1978).

In 1983 it was stated (Ilijanic 1983) that during the last 30 years great changes had taken place in the meadow vegetations of North Croatia. In the Drava region many former meadows had been turned into arable land, e.g. very interesting Molinietalia meadows. As a consequence of drainage works and intensive use, large areas had also been turned into Arrhenatherion vegetation, now being the most important meadow community. These meadows are poorer in species (Ilijanic, pers. comm.).

Apart from lowland grassland, grasslands situated in other regions (e.g. in mountains) are at least partly very well developed. Seckel (1991) mentions large areas of unfertilized meadows with a rich flora in the new national park of Kopaonik in Serbia. The same applies to the region around the Golija national park.

Czechoslovakia

According to Grimmett and Jones (1989), 54 per cent of the country is farmed, 25 per cent of this (1,600,000 ha) being grassland.

E. Balátová (pers. comm.) estimates that about 10-15 per cent is semi-natural grassland. Mesobromion still occurs on scattered sites, while *Arrhenatheretum* has become rare. Molinietalia meadows are most common in the river floodplains, where they may extend to over 100 ha.

The decline continues, due to drainage, conversion to arable and abandonment. Calthion and Molinion are threatened by drainage. Cnidion and Veronico-Lysimachion meadows disappear gradually, owing to the regulation of rivers and brooks. Alopecurion also has little chance of survival.

Though a great number of grasslands have been protected, these are partly threatened by lack of management. However, while the situation described is not at all favourable, it seems to be less serious than in the former GDR.

Hungary

The area of grassland is about 1.3-1.4 million ha (IUCN 1990, Grimmett and Jones 1989). This is 14-15 per cent of the country, whereas 58 per cent is arable land.

Of this area, 500,000 ha are extensively managed, 200,000-250,000 ha being of high conservation interest. The areas concerned, however, are scattered and fragmented (IUCN 1990).

Márkus (this volume) also gives national figures. The national area of grasslands is 1.2 million ha, 121,413 being protected (24 per cent of all protected areas). In his inventory it is recommended to protect at least another 15,000 ha of grassland. Further, he estimates another 250,000 ha is being used extensively.

Márkus states that the majority of the large continuous grasslands has been lost during the last 40 years and that grassland areas are still shrinking at an alarming rate. Nevertheless, steppic habitats of over 500 ha each still cover 154,400 ha. The 139,500 ha of sites, being protected in one way or another, embrace 15 sites of 1,700-52,000 ha. Management (especially under-grazing) is a great problem and only 15,000 ha are managed by nature conservation authorities.

All the following information was derived from IUCN (1990): Both original steppe climax vegetation and secondary grasslands occur; *puszta* are mostly semi-natural, secondary grasslands. Three main groups of grasslands are described for the Great Plain: loess grassland, chalk-sand grassland and alkaline grassland. Loess grassland used to be the only treeless climax vegetation. It has almost completely disappeared through ploughing. One association is mentioned. Chalk-sand grassland still occurs, though it has suffered both from afforestation and agricultural development. The majority of the remaining grassland is protected. Three associations are listed. Alkaline grasslands have suffered from various threats but no data are given on the present status. Seven associations are listed.

There is a wide variety of current threats, including over-grazing, under-grazing, fire, fertlization, and conversion to a number of other types of land use, including military manoeuvres.

Romania
Very little information was available for this survey, save some general information (Grimmett and Jones 1990). Romania holds 4.4 million ha of grassland, 29 per cent of the agricultural area, which in its turn covers 63 per cent of the land surface. There are remnants of steppe-woodland and steppe and important areas of riverine meadows and forests along the main rivers.

European USSR
Some general information was available (IUCN 1990). European Russia has 55.6 million ha of lowland grassland, the Ukraine 6.6 million ha, Byelorussia 3.3 million ha, the Baltic republics 2.2 million ha and Moldavia 0.3 million ha.

The total area of protected grasslands in nature reserves is 77,000 ha, in national parks 77,000 ha and in nature sanctuaries 1,837,000 ha. Of this last category, 1.7 million ha lie in Russia and 146,000 ha in Byelorussia whereas the Ukraine has only some 2,200 ha.

It is concluded that the protection of the full range of grassland types is inadequate. Flood-meadow grasslands are particularly under-represented and natural grasslands are poorly protected.

G. Regnéll (pers. comm.) reports that very detailed botanical information is available about Estland and that large areas of non-fertilized semi-natural grasslands still exist there.

V. Flint (this volume) makes it clear that the great majority of steppe habitats in the European part of USSR has been converted to arable land. As a result great bustards now breed only on arable land.

Grimmett and Jones (1989) mention some nature reserves, important for grassland habitats. The biggest is Askania-Nova, in the Ukraine. The Biosphere Reserve covers 33,307 ha, of which 1,500 ha are under total protection. The large Zavidovo reserve (RSFSR), covering 125,000 ha, includes 17,000 ha of meadows. The Central Black Earth Biosphere Reserve (RSFSR), a wooded steppe area, covers 4,800 ha.

Davydova and Koshevoi (1989) give a survey of nature reserves in the whole of the USSR by habitat type. For steppes and forest steppes 16 nature reserves are mentioned. Moreover, three important Biosphere Reserves are listed: the above-mentioned Askania-Nova, Central Black Earth and Chernomorsky. The latter being a steppe area near the Black Sea, covering 71,900 ha.

Finland, Bulgaria, Albania, Turkey
No data were available for this survey, but Sancar Baris (this volume) has highlighted the great importance of Turkey (Anatolia) for dry grasslands.

CONCLUSIONS AND RECOMMENDATIONS

1. Semi-natural grasslands are very important for the conservation of the European flora and fauna. In western and central Europe there are few primary habitats remaining that can maintain these communities.

2. A very great part of the former area of semi-natural grasslands in Europe has been destroyed in this century. In large areas often only one to a few per cent of the grassland area is still rich in species. Destruction goes on.

As the vegetation of semi-natural grasslands can withstand only very little fertiliser application, far lower than that which present day agriculture requires, protecting the habitat and flora is generally far more difficult than protecting grassland bird species.

3. In the past, fertilization, ploughing and drainage were important causes of loss. Now abandonment (including afforestation) of the last remaining species-rich grasslands is a great threat too, if not the greatest. Abandonment can even take place within farms when parts are intensified and others become superfluous.

There are a number of other threats like sand, gravel and clay extraction; acidic, nitrogen-rich precipitation etc., but none has had such a large-scale impact as agricultural intensification.

4. Important strongholds of remaining semi-natural grasslands include parts of Ireland, rough grazings in Great Britain, several river valleys and secondary mountain chains in France, parts of southern Germany, river flood plains in eastern

Europe, e.g. in Poland and Yugoslavia (where good grasslands also exist in mountainous areas) and steppe habitats in Hungary. In the USSR, important resources may be present, but little information was available for this survey. The same can be said about southern Europe.

5. For long-term protection of grassland habitats it is necessary to achieve durable systems of management, of which "extensive" grazing of large areas seems to be one of the best possibilities. A greater variety of management techniques can be applied to nature reserves, depending on their size.

Such management systems can be administered by e.g. nature conservation bodies, regional parks and national parks. Livestock for grazing can be either a herd of "primitive" breeds (e.g. Heck cattle, Highland cattle, certain French races of cattle, Konik horses and certain breeds of sheep) or herds of privately owned cattle from the surrounding areas, changed on a yearly basis.

Such systems also permit the development of complexes of different natural elements like grasslands, scattered trees and (plots of) scrub and woodland, where it is not essential for bird populations to keep the areas completely open. Both systems of management do exist already, e.g. in the Netherlands.

Durable systems of botanically rich grasslands (mostly demanding a far lower input, or none at all, of fertiliser than many ornithologically important grasslands) in the form of private agriculture seem to be possible only if there is a good market for products, produced according to ecological criteria.

Short-term solutions, resulting in any sort of abatement of negative trends, giving more time to work for durable solutions, must be welcomed and can be attained by different measures or combinations of them, like the support for less favoured areas or management agreements systems and regional production quotas. Moreover, if this can result in durable systems it must certainly be welcomed, as large areas could be protected in this way. However, immediate action is necessary if we want to use the existing opportunities.

6. In several countries with the most important remaining semi-natural grasslands, relatively little has been done in the field of conservation measures for grassland habitats. It is here that great benefits can be attained if measures are taken quickly. In EC countries, implementation of existing and future EC regulations offer good "tools", whereas for east European countries western support must be recommended.

7. In EC countries, the application of Directive EC/797/85/Article 19, making management agreements possible, must be strongly recommended. The objective must be to prevent abandonment and intensification at the same time. The equilibrium is very vulnerable!

Other measures in favour of such areas are: stimulation of the production of regional products (e.g. regional cheese brands) and further adaptation of the production of them to ecological criteria; introduction of regional instead of national production quota in order to protect such areas (recent plea of the Stichting Natuur en Milieu, D. Logemann, pers. comm.).

There needs to be investigation into how disproportionate afforestation of low input areas (marginal land from agricultural viewpoint), can be avoided. For nature conservation and for the reduction of production surpluses, it would be better to afforest more productive areas.

8. The creation of nature reserves, both small and large, is important for species that cannot survive in the present day form of agriculture. In the Netherlands for example, with a very intensive agriculture, management agreements are not suitable to protect semi-natural grasslands but they are for the protection of birds.

Small nature reserves can be the basis for future colonisation of the surroundings, e.g. in the framework of restoration projects. Large reserves give a unique chance for the development of more complete ecosystems and they can function as special genetic reservoirs because of their species diversity and vital populations. Nature reserves should be linked to form a biogenetic network. Concerning dry grasslands, a proposal for such a network was presented to the Council of Europe in 1982 (Géhu 1984).

9. All forms of agricultural development like land consolidation and drainage, and also regulation of rivers still present threats to semi-natural grasslands, especially in the lowlands. Such schemes should be critically evaluated before being allowed to proceed. This does not apply only to national projects, but also to EC-funded projects and multilateral and bilateral aid to east European countries.

10. In eastern Europe there are some very important regions from the viewpoint of grassland conservation that merit international attention. From the data which were available, it appears that river valleys in Poland and Yugoslavia, extensive (large) meadow areas in the mountainous parts of Yugoslavia and steppe habitats in Hungary belong to this category. Immediate threats to these areas are amongst others: intensification (including drainage), diking (flood plains, resulting in changes on both sides of the dike), abandonment of botanically rich but little productive land and privatisation of traditionally managed common land, or conversion of them into intensively used collective farms.

Some of the most important options for Eastern Europe seem to be:

Where privatisation of collective farms takes place, this gives a unique opportunity to form a coherent network of large nature reserves. Though the present quality may be low, prospects for a certain degree of restoration do certainly exist, not in the least from an ornithological viewpoint. In the former GDR, such opportunities are presently being used to full advantage (Litzbarski pers. comm.).

Ancient common pastures should be maintained as far as possible because of their unique ecological situation. Grazing rights should be retained.

In certain parts of eastern Europe, the gap between present-day agriculture and the requirements of "ecological" or "organic" farming is much smaller than in western Europe, provided pollution is not a problem. Export opportunities to the EC market for such products could be a stimulus for this sort of farming, providing this would not affect the position of similar farming in the west.

Further recommendations, as mentioned under 5-9 above, are certainly important here too. In cases where either intensification or abandonment are immediate threats, action is very urgent.

ACKNOWLEDGEMENTS

I acknowledge the help given by a great number of people and institutions. The Directie Natuur-, Milieu- en Faunabeheer gave me three weeks for travel to Poland and other activities, as well as some other facilities. Miss M. Smolders and Mr. H. Veerman (Staatsbosbeheer) helped greatly with general literature research. Mr. J. Vink (Staatsbosbeheer) put his grassland archive at my disposal. General information was also received from Dr. J. Willems, Drs. ing. P. H. Veen and Drs. D. Logemann. Very useful comments on the draft were received from Dr. A. M. O'Sullivan. The other sources are mentioned below, but this does not mean that they are responsible for the text concerning their country.

Useful information was received from: Netherlands: Dr. F.J. van Zadelhoff, ir. W. Geraedts, Dr. K.V. Sykora. Great Britain: Mr. R. Fuller, Mr. D. Glaves. Ireland: Dr. J. Lee (also other countries), Dr. A.M. O'Sullivan, Dr C. Ó'Críodáin. Germany: Dr. F. Ziesemer, Prof. Dr. Drs.h.c. H. Ellenberg. Dr. U. Bohn, Dr. agr. G. Briemle, Prof. Dr. O. Wilmanns, Dr. S. Klotz, Dipl. -Ing.(TV) E. Wenisch, Dr. H. Schulzke, Dr. D.M. Woike, Dipl. Ing. E. Schulte, Dr. J. Heringer, Dr. B. Schall. France: Dr. J. M. Royer, Prof. Dr. R. Delpech, Mr. H. Maurin, Mr. Bournerias, Mr. J.R. Zimmermann, Mr. M. Forestier, Mr. J. C. Lefeuvre, Mrs. N. Yavercovski. Switzerland: Dr. T. Dalang. Austria: Dr. A. Polatschek, Prof. Dr. G. Grabherr and Dr. L. Schratt. Belgium: Dr. E. Kuijken. Norway: Dr. M.H. Losvik. Sweden: Mr. K. Ekeland, Dr. G. Regnéll, Prof. Dr. E. van der Maarel. Poland: Prof. dr. hab. D. Fijałkowski, dr. E. Chojnacka-Fijałkowska and Doc. Dr. hab. M.T. Rogalski. Czechoslovakia: Dr. E. Balatova. Yugoslavia: prof. dr. Lj. Ilijanic, Drs. F. Vera.

REFERENCES

Bakker, J.J., van Dessel, B. and van Zadelhoff, F.J. (1989). *Natuurwaardenkaart 1988*. SDU-uitgeverij, Den Haag.

Bakker, J.P. (1982). Veranderingen in vochtige graslandvegetaties onder invloed van hooien zon der bemesting. *Vakblad voor Biologen* 62: 43-48.

Bakker, J.P. (1985). Hooien zonder bemesting: hoe langer hoe schraler? *De Levende Natuur* 86: 149-153.

Baldock, D. (1990). *Agriculture and habitat loss in Europe*. WWF-International CAP Discussion Paper No. 3.

Baldock, D. Hermans, B.P.G.M., Kelly, P. and Mermet, L. (1984). *Wetland drainage in Europe*. Institute for European Environmental Policy/The International Institute for Environment and Development. London.

Baldock, D. and Long, A. (1987). *The Mediterranean Environment under Pressure: the influence of the CAP on Spain and Portugal and the "IMPs" in France, Greece and Italy*. Report to WWF.

Barbero, M., Loisel, R. and Quezel, P. (1984). Les pelouses calcaires du sud-est de la France: facteurs de pression et problèmes de protection. In: Géhu, J.M. (Ed.) *La végétation des pelouses calcaires, Strasbourg 1982*, pp. 185-193. Cramer, Vaduz.

Biondi, E. and Blasi, C. (1984). Les pelouses seches calcaires à *Bromus erectus* de l'Apennin central et meridional (Italie). In: Géhu, J.M. (Ed.) *La végétation des pelouses calcaires, Strasbourg 1982*, pp. 195-200. Cramer, Vaduz.

Böttcher, W. and Schlüter, H. (1989). Vegetationsveränderung im Grünland einer Flussaue des Sächsischen Hügellandes durch Nutzungsintensivierung. *Flora* 182: 385-418.

Briemle, G. (1988). Ist eine Schafbeweidung von Magerrasen der Schwäbischen Alb notwendig? *Veröff. Naturschutz Landschaftspflege Baden-Württemberg* 63: 51-67.

Briemle, G. (1990). Über die Wirkung mineralischer Düngung auf die Vegetation einer Enzian-Magerwiese der Schwäbischen Alb. *Natur und Landschaft* 65: 315-319.

Broggi, M. (1989). *Mindestbedarf an naturnahen Flächen in der Kulturlandschaft*. Bericht 31 der NFP "Boden", Liebefeld-Bern.

Bush, M.B. and Flenley, J.R. (1987). The age of British chalk grassland. *Nature* 329: 434-436.

Centraal Bureau voor de Statistiek (1989). *Algemene milieustatistiek 1989*. SDU-uitgeverij, Den Haag.

Corporaal, A. (1990). *De Loire en de Kievitsbloem*. Ministerie van Landbouw, Natuurbeheer en Visserij, Den Haag.

Committee of endangered plants (1982). *List of rare, threatened and endemic plants in Europe*. Council of Europe, Strasbourg.

Curtis, T.G.F. and McGough, H.N. (1988). *The Irish Red Data Book. I Vascular Plants*. Wildlife Service, Dublin.

Davydova, M. and Koshevoi, V. (1989). *Nature Reserves in the USSR*. Progress Publishers, Moscow.

de Juana, E. Santos, T., Suarez, F. and Telleria, J.L. (1988). Status and conservation of steppe birds and their habitats in Spain. In: Goriup, P.D. (Ed.) *Ecology and conservation of grassland birds*, ICBP Technical Publication No. 7, Cambridge, pp. 113-124.

Delpech, R. (1975). *Resumé de la thèse de doctorat es sciences naturelles*. Laboratoire de Taxonomie végétale expérimentale et numérique de l'Université de Paris-Sud, Paris.

Dijk, G. van (1990). *Notes on grassland conservation in Poland* (draft).

Drachenfels, O., Mey, H. and Miotk, P. (1984). *Naturschutzatlas Niedersachsen*. Niedersächsisches Landesverwaltungsamt, Fachbehörde für Naturschuutz, Hannover.

Drachenfels, O. von and Mey, H. (1988). *Erfassung der für den Naturschutz wertvollen Bereiche in Niedersachsen*. Informationsdienst Naturschutz Niedersachsen 4/88.

Dierssen, K., von Glahn, H., Härdtle, W., Höper, H., Mierwald, U., Schrautzer, J. and Wolf, A. (1988). *Rote Liste der Pflanzengesellschaften Schleswig-Holsteins*. Schriftenreihe des Landesamtes für Naturschutz und Landschaftspflege Schleswig-Holstein, Heft 6. Kiel.

Duffey, E., Morris, M.G., Sheail, J., Ward, L.K., Wells, D. A. and Wells, T.C.E. (1974). *Grassland Ecology and Wildlife Management*. Chapman and Hall, London.

Duvigneaud, J. (1984). Le pré à *Bromus erectus* et *Thalictrum minus* subsp. *majus* de la plaine alluviale de la Moselle. In: Géhu, J.M. (Ed.), *La végétation des pelouses calcaires, Strasbourg 1982*, pp. 269-280. Cramer, Vaduz.

Ellenberg, H. (1986). *Végétation Mitteleuropas mit den Alpen*. Ulmer, Stuttgart.

Fuller, R.M. (1987). The changing extent and conservation interest of lowland grasslands in England and Wales: A review of grasslands surveys 1930-84. *Biological Conservation* 40: 281-300.

Fijałkowski, D. and Chojnacka-Fijałkowska, E. (1990). Zbiorowiska z klas Phragmitetea, Molinio-Arrhenatheretea i Scheuzerio-Caricetea fuscae w makroregione Lubelskim. *Polish Agricultural Annual. Seria D-Monografie-Tom* 217, Warszawa.

Ganzert, C. and Pfadenhauer, J. (1989). Végétation und Nutzung des Grünlandes am Dümmer. *Naturschutz und Landschaftspflege in Niedersachsen* 16. Niedersächsisches Landesverwaltungsamt-Fachbehörde fur Naturschutz.

Gaspar, A.M., de Sequeira, E.M. and Dordio, A.M. (1988). National Report Portugal. *In:* Park,J.R. (Ed.) *Environmental Management in Agriculture, European Perspectives,* 75-82. Belhaven Press, London and New York.

Geiser, R. (1983). Die Tierwelt der Weidelandschaften. *In: Akademie für Naturschutz und Landschaftspflege.* Schutz von Trockenbiotopen: Trockenrasen, Triften und Hutungen. Laufener Seminarbeiträge 6/83.

Géhu, J.M. (Ed.) (1978a). *La végétation des prairies inondables.* Lille 1976. Cramer, Vaduz.

Géhu, J.M. (Ed.) (1978b). *La végétation des pelouses sèches à thérophytes.* Lille 1977. Cramer, Vaduz.

Géhu, J.M. (Ed.) (1984). *La végétation des pelouses calcaires, Strasbourg 1982.* Cramer, Vaduz.

Grimmett, R.F.A. and Jones, T.A. (1989). *Important Bird Areas in Europe.* ICBP Technical Publication No. 9. ICBP, Cambridge.

Gustawsson, K.A. (1976). *Ängen och hagen.* Almqvist and Wiksell International, Stockholm.

Hedinger, C. (1984). Lebensraum Trockenrasen. *Schweizer Naturschutz* 4/84: 1-25.

Hillier, S.H., Walton, D.W.H. and Wells, D.A. (Eds) (1990). *Calcareous Grasslands-Ecology and Management. Proceedings of a joint BES/NCC symposium, Sheffield, 1987.* Bluntisham Books, Bluntisham, Huntingdon.

Holzner, W., Horvatic, E., Köllner, E., Köppl, W., Pokorny, M., Scharfetter, E., Schramayr, G. and Strudl, M. (1986). *Österreichischer Trockenrasenkatalog.* Grüne Reihe des Bundesministeriums für Gesundheit und Umweltschutz Band 6.

Hopkins, A. and Wainwright, J. (1989). Changes in botanical composition and agricultural management of enclosed grassland in upland areas of England and Wales, 1970-86, and some conservation implications. *Biological Conservation* 47: 219-235.

Hopkins, A., Wainwright, J., Murray, P.J., Bowling, P.J. and Webb, M. (1988). 1986 survey of upland grassland in England and Wales. *Grass and Forage Science* 43: 185-198.

Ilijanic, L. (1973). Allgemeiner Überblick über die wechselfeuchten Niederungswiesen Jugoslaviens im Zusammenhang mit den klimatischen Verhältnissen. *Acta Bot. Ac. Sci. Hungaricae* 19: 165-179.

Ilijanic, L. (1988). Über die Grundwasserverhältnisse unter einigen Wiesengesellschaften in Nordwestkroatien. *Acta Bot. Croat.* 47: 41-61.

Ilijanic, L. and Segulja, N. (1978). Zur pflanzensoziologischen Gliederung der Glatthaferwiesen Nordostkroatiens. *Acta Bot. Croat.* 37: 95-105.

Ilijanic, L. and Segulja, N. (1983). Phytozönologische und Ökologische Untersuchungen der Glatthaferwiesen in der Podravina (Nordkroatien). *Acta Bot. Croat.* 42: 63-82.

IUCN (1990). *The lowland grasslands of Eastern Europe, A Survey, with selected country case studies.* IUCN, East-European Programme, Cambridge.

Keymer, R.J. and Leach, S.J. (1990). Calcareous grasslands-a limited resource in Britain. *In:* Hillier, S.H., Walton, D.W.H. and Wells, D.A. (Eds) *Calcareous Grasslands-Ecology and Management. BES/NCC-symposium, Sheffield, 1987.* Bluntisham Books, Bluntisham, Huntingdon.

Klafs, G., Stübs, J., Grempe, G., Lambert, K., Nehls, H.W. and Starke, W. (1979). *Die Vogelwelt Mecklenburgs.* VEB Gustav Fischer Verlag, Jena.

Klein, A. and Keller, H. (1983). Trockenstandorte und Bewirtschaftungsbeiträge. Bundesamt für Forstwesen Bern.

Korneck, D. and Sukopp, H. (1988). *Rote Liste der in der Bundesrepublik Deutschland ausgestorbenen, verschollenen und gefährdeten Farn- und Blütenpflanzen und ihre Auswertung für den Arten- und Biotopschutz. Schriftenreihe Végétationskunde 19.* Bundesforschungsanstalt für Naturschutz und Landschaftsökologie, Bonn-Bad Godesberg.

Kruk, M. (1991). *Farming and nature conservation in Britain.* Dept. of Env. Biology. State University of Leiden.

Kuijken, E. (1988). Applied ecological research on the conservation of wet grasslands in relation to agricultural use in Flanders. *In:* Park, J.R. (Ed) *Environmental management in agriculture,* 207-215. Belhaven Press, London and New York.

Landesanstalt für Ökologie, Landschaftsentwicklung und Forstplanung (Nordrhein-Westfalen) (1988). Mittelgebirgsprogramm NRW. Umweltschutz und Landwirtschaft 4.

Lee,J. (1990). *Land use trends and factors influencing change in future land use in EC-12.* European Agrarian Youth Congress, Groningen.

Logemann, D. (1986). *Het EG-landbouwbeleid en het milieu.* Stichting Natuur en Milieu, Utrecht.

Losvik, M.H. (1988). Phytosociology and ecology of old hay meadows in Hordaland, western Norway in relation to management. *Végétatio* 78: 157-187.

Lübbe, E. (1988). National Report West Germany. *In:* Park, J.R. (Ed) *Environmental management in agriculture,* 83-94. Belhaven Press, London and New York.

Maarel, E. van der (1988). Floristic diversity and guild structure in the grasslands of Öland's Stora Alvar. *In:* Sjögren, E. (Ed.) *Plant cover on the limestone Alvar of Öland,* 53-65. Almqvist and Wiksell International, Stockholm.

Mathers, M. and Woods, A. (1989). Making the most of Environmentally Sensitive Areas. *RSPB Conservation Review* 3: 50-55.

Ministry of Agriculture, Fisheries and Food (MAFF) (1989). *Environmentally Sensitive Areas.* HMSO London.

Mulqueen, J. (1988). National Report Ireland. *In:* Park, J.R. (Ed) *Environmental management in agriculture,* 41-48. Belhaven Press, London and New York.

Murphy, W.E. and O'Keefe, W F. (1985). Fertiliser use survey. *In:* Fertiliser Association of Ireland, Publication No. 28.

Naturvårdsverket (1987). *Inventering av ängs- och hagsmarker.* Handbok. Solna.

Noirfalise, A. (1988). National Report Belgium. *In:* Park, J.R. (Ed.) *Environmental management in agriculture,* 37-40. Belhaven Press, London and New York.

Noirfalise, A. and Dethioux, M., (1984). Les pelouses calcaires de la Belgique et leur protection. *In:* Géhu, J.M. (Ed.) *La végétation des pelouses calcaires, Strasbourg 1982,* pp. 201-218. Cramer, Vaduz.

Oomes, M.J.M. (1983). De invloed van lage bemestingsgiften op de botanische samenstelling van grasland onder gebruiksbeperkingen. *Bosbouwvoorlichting* 22: 5-8.

O'Sullivan, A.M. (1982). The lowland grasslands of Ireland. *Journal of Life Sciences* 3(1/2): 131-142.

Ouborg, N.J., (1988). Genetische verarming: de problematiek van het beheer van kleine plantenpopulaties. *De Levende Natuur* 89: 7-13.

Pautou, G. and J. Girel (1984). Genèse, évolution et disparition des pelouses calcaires dans la plaine alluviale du Rhône entre Genève et Lyon. *In:* Géhu, J.M. *La végétation des pelouses calcaires, Strasbourg 1982,* pp. 239-242. Cramer, Vaduz.

Pottier-Alapetite, G., (1943). *Recherches phytosociologiques et historiques sur la végétation du Jura central et sur les origines de la flore jurassienne.* SIGMA communication 81, Montpellier.

Preiß, H. and Haxel, H. (1982). Seminarergebnis. *In:* Waldweide und Naturschutz. Laufener Seminarbeiträge 9/82. Akademie für Naturschutz und Landschaftspflege.

Ratcliffe, D.A. (1984). Post-medieval and recent changes in British végétation: the culmination of human influence. *The New Phytologist* 98: 73-100.

Riely, J.O. and Page, S.E. (1990). *Ecology of plant communities, a phytosociological account of the British végétation.* Longman Scientific and Technical/Wiley and Sons, New York.

Royer, J.M. (1987). Les pelouses des Festuca-Brometea: d'un

exemple régional à une vision eurosibérienne. *Etude phytosociologique et phytogéographique.* Thèse, 3 vol. Besançon.

Royer, J.M. (year?). Essai de synthèse sur les groupements végétaux de pelouses, éboulis et rochers de Bourgogne et Champagne méridionale. *Ann. Sc. Univ. Besançon,* 3ième série, 13: 157-316.

Royer, J. M. and Didier, B. (1982). Etude phytosociologique des pairies alluviales inondables du Bassin de la Voire (Champagne humide, France). *Bulletin de la Société de Sciences Naturelles et d'Archéologie de la Haute Marne,* Tome XXI, fasc. 17.

Ruiz Perez, M. (1986). *Sustainable food production in the Spanish dehesa.* Publication of the Food Energy Nexus Programme. United Nations University.

Ruiz Perez, M. and Groome, H. (1986) Spanish agriculture in the EEC: a process of marginalization and ecological disaster? *F.F.S.P.N. Rencontres Internationales de Toulouse Agricultre-Environment 1986:* 456-461.

Rutschke, E., Libbert, W., Litzbarski, H., Schmidt, A. and Schummer, R. (1983). *Die Vogelwelt Brandenburgs.* VEB Gustav Fischer Verlag, Jena.

Schotsman, N. (1988). *Onbemest Grasland in Friesland, Hydrologie, typologie en toekomst.* Provincie Friesland, Hoofdgroep Ruimtelijke Ordening, Leeuwarden.

Schneider-Jacoby, M. (1990). *Projekt: Save-Aven.* Stiftung Europäisches Naturerbe, Radolfzell.

Seckel, B.J. (1991). Orchideeën in het Nationale Park Kopaonik in Joegoslavië. *Natura* 88(2): 37-39.

Shimwell, D.W. (1971). Festuco-Brometea Br.-Bl. and R.Tx 1943 in the British Isles: the phytogeography and phytosociology of limestone grasslands. *Végétatio* 23: 1-28 (I) and 29-60 (II).

Sissingh, G. (1978). Le *Cirsio-Molinietum* Sissingh et De Vries (1942) 1946 dans les Pays-Bas. *In:* Géhu, J.M. (Ed.) *La végétation des prairies inondables,* Lille 1976: 289-301. Cramer, Vaduz.

Speybroeck, D. van, (1984). Observations phytosociologiques sur les pelouses calcaires de la Vallée de la Somme. *In:* Géhu J.M., (Ed.) *La végétation des pelouses calcaires, Strasbourg 1982,* pp. 105-115. Cramer, Vaduz.

Stanley, D. (1989). *Eastern Europe on a shoestring.* Lonely Planet Publications, Hawthorn (Australia).

Stott, P.A. (1971). A *Mesobrometum* referable to the sub-association *Mesobrometum seslerio polygaletosum* Tüxen described for the Somme Valley. *Végétatio* 23: 61-70.

Sykora, K.V. and Zonderwijk, P. (1986). Kleurrijke dijkbeemden: hoe lang nog? *Waterschapsbelangen* 71: 247-253.

Westhoff, V. and den Held, A.J. (1969). *Plantengemeenschappen in Nederland.* Thieme, Zutphen.

Westhoff, V. and Weeda, E. (1984). De achteruitgang van de Nederlandse flora sinds het begin van deze eeuw. *Natuur en Milieu* 8: 8-17.

Wilkinson, (1987). Agriculture and nature conservation. *Naturopa* 56: 5-7.

Willems, J.H. (1980). *Limestone grasslands in north-west Europe.* Thesis. University of Utrecht.

Willems, J.H. (1987). Ons Krijtland Zuid-Limburg VI, Kalkgrasland in Zuid-Limburg. Wetenschappelijke Mededeling KNNV 184.

Willems, J.H. (1990). Calcareous grasslands in continental Europe. *In:* Hillier, S.H., Walton, D.H.W. and Wells, D.A. *Calcareous grasslands - Ecology and management. BES/NCC-symposium, Sheffield, 1987.* Bluntisham Books, Bluntisham, Huntingdon.

Wilmanns, O. (1988). Können Trockenrasen derzeit trotz Immissionen überleben? - Eine kritische Analyse des *Xerobrometum* im Kaiserstuhl. *Carolinea* 46: 5-16.

Wilmanns, O. (1989). Zur Entwicklung von Trespenrasen im letzten halben Jahrhundert: Einblick-Ausblick-Rückblick, das Beispiel des Kaiserstuhls. Düsseldorffer Geobot. *Kolloq.* 6: 3-17.

Wilmanns, O. (1989). Dynamik und Schutz von Pflanzengesellschaften im Kaiserstuhl/Südbaden. *In: Naturschutz- und Umweltpolitik als Herausforderung.* Hannover.

Wolkinger, F. and Plank, S. (1981). *Dry grasslands of Europe.* Council of Europe, Strasbourg.

Woike, M. (1988a). *Die Bedeutung des Grünlandes im Mittelgebirge für den Naturschutz sowie Möglichkeiten seiner Erhaltung.* Seminarberichte des Naturschutzzentrums Nordrhein-Westfalen Band 2, Heft 4, 1988.

Woike, M. (1988b). *Grünlandprogramme in Nordrhein-Westfalen.* Jahrbuch für Naturschutz und Landschaftspflege, Band 41: 105-122. Bonn.

Ziesemer, F. (1989). Entwicklung und erste Ergebnisse des Extensivierungsprogrammes in Schleswig-Holstein. ICBP, German Section, *Bericht* 28: 77-85.

THE STATUS OF LOWLAND DRY GRASSLAND BIRDS IN EUROPE

Graham M Tucker

International Council for Bird Preservation, 32 Cambridge Road, Girton, Cambridge CB3 0PJ, UK

ABSTRACT

The status of European lowland dry grassland birds is reviewed in terms of their current population sizes, trends, conservation priorities and protection through existing international conservation measures. The review used currently published information on population sizes and trends and produced a preliminary list of 27 species of dry grassland birds identified as being of conservation concern, 93 per cent of which are declining. Of these species of conservation concern four are globally threatened; lesser kestrel *Falco naumanni*, little bustard *Tetrax tetrax*, great bustard *Otis tarda* and sociable plover *Chettusia gregaria*. All of these globally threatened species are declining.

Most populations of European dry grassland birds are concentrated in Spain and the Soviet Union. The importance of these areas probably reflects both the extent of grassland in these countries and the relatively low intensity agriculture in these regions.

Despite some limitations of the currently available data, it is clear that existing international conservation measures have had little effect on grassland birds. There are currently few protected areas of relevance to grassland birds. Their dispersed distribution necessitates a broader conservation approach to habitats through land-use policies which take account of wildlife requirements.

INTRODUCTION

Lowland dry grassland has an avifauna which is greatly threatened by habitat loss and degradation largely through agricultural expansion and intensification (Goriup and Batten 1990) although also in some cases agricultural abandonment (Rodríguez and de Juana, this volume). However, other than for those species which are globally threatened the current status of most dry grassland birds is poorly known and little attempt has been made to identify those species in need of urgent conservation measures. Using currently available information this paper attempts to review the population status and trends of European dry grassland birds in order to:

i) identify which species are of conservation concern;
ii) establish the efficiency of international conservation measures;
iii) establish the locations of European populations of globally threatened birds.

THE DEFINITION OF LOWLAND DRY GRASSLAND BIRDS

Lowland dry grassland

For the purpose of this review lowland dry grassland is defined as wholly or virtually treeless lowland plains or undulating ground dominated by non-ericaceous species averaging less than 1 metre high (Goriup and Batten 1990). Grassland which is irrigated or waterlogged at some point during the year is excluded. In Europe this "steppic" habitat is generally the result of human activities on the landscape; initially the destruction of woodland and subsequently the effects of grazing. Now these remaining grasslands tend to be either outposts of the vast Eurasian primary steppes dominated by drought-resistant *Festuca* and *Stipa* species, or sub-Mediterranean shrubby grasslands (Wolkinger and Plank 1981, Goriup 1988). Other more restricted areas of dry grassland includes those formed on thin soils over chalk or limestone as in the UK and Ireland, or on saline soils near the coasts or occasionally inland as in central Turkey and Spain (Goriup and Batten 1990).

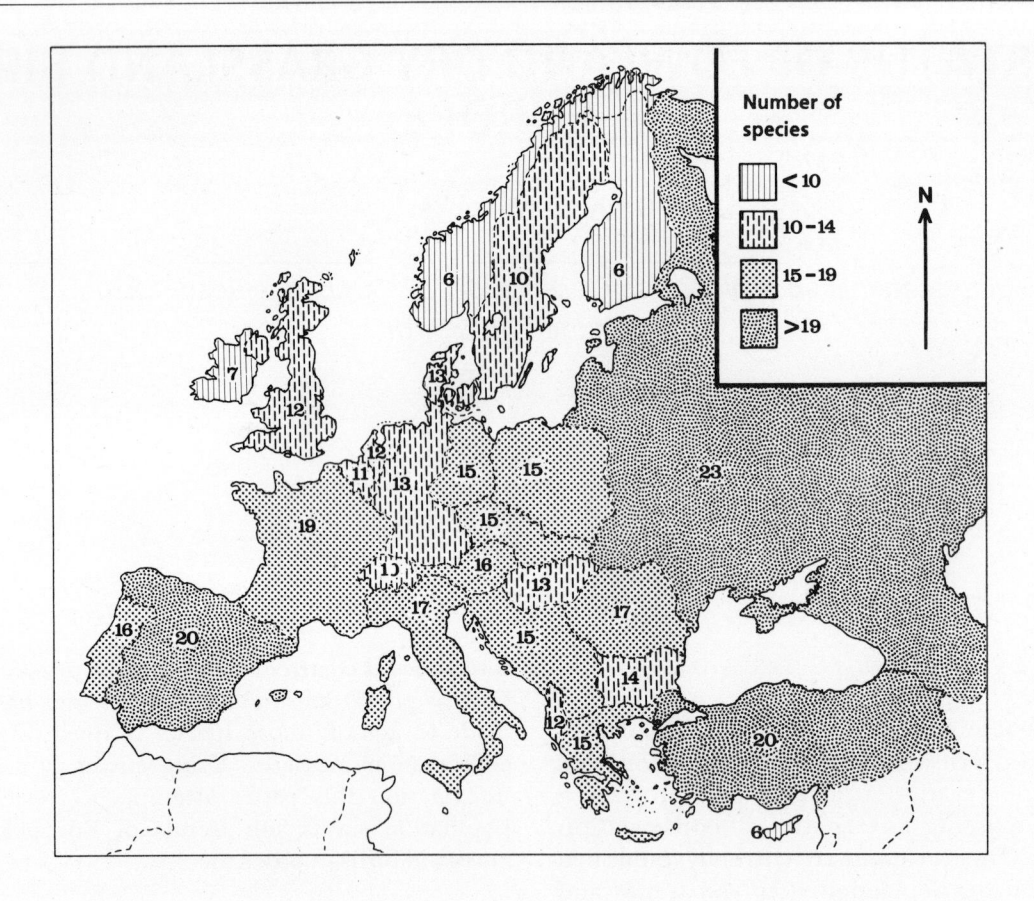

Figure 1: The number of European dry grassland species of conservation concern in each country

The avifauna of dry grassland

Dry grassland birds are here defined as those that either predominantly nest within dry grassland habitats (as defined above) or predominantly feed within these at some stage of their annual cycle (Table 1). Unavoidably, this list is arbitrary regarding those species that use other habitats as well as dry grassland, and especially regarding those that inhabit areas that verge on wet grassland or semi-desert areas. Furthermore, although these listed birds are all typical of dry grassland habitats, they do not constitute a distinct community. Indeed different European dry grassland areas differ substantially in their species composition as a result of the wide geographical area, disparate climates and subtle but important differences in vegetation composition and structure (e.g. de Juana *et al.* 1988, Petretti 1988). Furthermore, only a few (i.e. sociable plover, black lark *Melanocorypha yeltoniensis* and isabelline wheatear *Oenanthe isabellina* are restricted at some point in their annual cycle to dry grassland. Others are able to adapt to steppic habitats which have been further modified by human intervention, thus occurring on improved grassland or cultivated areas, termed "pseudo steppes" by Goriup (1988), where agricultural methods are not too intensive. Of the other habitats used by grassland species, these include desert and semi-desert e.g. in the case of long-legged buzzard *Buteo rufinus*, saker *Falco cherrug*, black-bellied sandgrouse *Pterocles orientalis*, pin-tailed sandgrouse *Pterocles alchata* and lesser short-toed lark *Calandrella rufescens*; wet grass habitats e.g. in the case of crane *Grus grus*, short-eared owl *Asio flammeus* and meadow pipit *Anthus pratensis*, and more intensive agricultural areas, e.g. in the case of red-legged partridge *Alectoris rufa*, grey partridge *Perdix perdix*, skylark *Alauda arvensis* and rook *Corvus frugilegus*.

Clearly those species with the most restricted habitat requirements are most vulnerable to the loss of natural grassland, whilst those which have adapted to more intensive agricultural habitats are unlikely to be threatened by habitat loss.

THE CONSERVATION STATUS OF DRY GRASSLAND BIRDS

Priorities for conservation

The need for the conservation of a species within a specific area should take account of two main factors. Firstly, the global distribution of the species, i.e. is the species merely on the edge of its range or is its core population within the area in question? Secondly, the risk to the population within the area, as assessed by its size, tendency towards decline or degree of restriction to a small number of potentially vulnerable sites. Therefore to identify which of the grassland bird species identified above are of conservation concern in Europe, data on these species has been collated to prioritise the species according to these

Table 1: Habitat use by dry grassland bird species in Europe

Definitions of habitat types are:

Dry grasslands: see text
Semi-improved agricultural grass: fertilized and/or reseeded grassland with irrigation in arid areas, but excluding heavily fertilized short-term grassland for silage
"Pseudo-steppe": cereal and fodder crops on traditionally cultivated land with low fertiliser and pesticide inputs

Habitat use codes:

B   Breeding habitat
W   Wintering habitat
P   Habitat used for feeding by passage birds which winter outside Europe

Birds are only classified as present if they regularly use the habitat for nesting and/or feeding

Species	Dry grassland	Semi-improved agricultural grassland	"pseudo-steppe"	Other habitats
Cattle egret *Bubulcus ibis*	BW	BW	BW	BW
White stork *Ciconia ciconia*	B	B	B	B
Hen harrier *Circus cyaneus*	W	W	W	BW
Pallid harrier *Circus macrourus*	BP		BP	P
Montagu's harrier *Circus pygargus*	BP		BP	BP
Long-legged buzzard *Buteo rufinus*	BW		?	BW
Steppe eagle *Aquila rapax*	B		?	B
Lesser kestrel *Falco naumanni*	BP	BP	BP	P
Lanner *Falco biarmicus*	BW			BW
Saker *Falco cherrug*	BW	?	BW	BW
Red-legged partridge *Alectoris rufa*	BW	BW	BW	BW
Grey partridge *Perdix perdix*	BW	BW	BW	
Quail *Coturnix coturnix*	B	B	B	
Crane *Grus grus*	B_a W		W	B
Demoiselle crane *Anthropoides virgo*	B_a W		BW	B
Little bustard *Tetrax tetrax*	BW	BW	BW	
Great bustard *Otis tarda*	BW		BW	
Stone curlew *Burhinus oedicnemus*	BW	BW	BW	BW
Collared pratincole *Glareola pratincola*	B	P	P	BP
Black-winged pratincole *Glareola nordmanni*	B	P	BP	BP
Sociable plover *Chettusia gregaria*	B			
Black-bellied sandgrouse *Pterocles orientalis*	BW		BW	BW
Pin-tailed sandgrouse *Pterocles alchata*	BW		BW_b	BW
Barn owl *Tyto alba*	BW	BW	BW	BW
Little owl *Athene noctua*	BW	BW	BW	BW
Short-eared owl *Asio flammeus*	BW	BW	W	BW
Dupont's lark *Chersophilus duponti*	B		W	BW
Calandra lark *Melanocorypha calandra*	BW	BW	BW	
White-winged lark *Melanocorypha leucoptera*	BW		W	BW
Black lark *Melanocorypha yeltoniensis*	BW			W
Short-toed lark *Calandrella brachydactyla*	B		B	B
Lesser short-toed lark *Calandrella rufescens*	BW			BW
Crested lark *Galerida cristata*	BW		BW	BW
Thekla lark *Galerida theklae*	BW		BW	BW
Skylark *Alauda arvensis*	BW	BW	BW	BW
Tawny pipit *Anthus campestris*	B	B	B	B
Meadow pipit *Anthus pratensis*	BW	BW	W	BW
Isabelline wheatear *Oenanthe isabellina*	B			
Wheatear *Oenanthe oenanthe*	B			B
Rook *Corvus frugilegus*	BW	BW	BW	BW
Rose-coloured starling *Sturnus roseus*	B			B
Corn bunting *Miliaria calandra*	BW	BW	BW	BW

a   Close to water
b   Very dry and infertile areas only

Table 2: Dry grassland bird populations in Europe, their population trends and conservation priorities

Species	European Population	% of world Population	Population trends	Conservation priority
Cattle egret *Bubulcus ibis*	58,000	<1-5	++	
White stork *Ciconia ciconia*	120,000	51-75	—	2
Hen harrier *Circus cyaneus*	5,300-10,000$_a$*	6-25	~	
Pallid harrier *Circus macrourus*	?	26-50	?	?
Montagu's harrier *Circus pygargus*	7,800-9,300$_a$*	51-75	—$_A$	2
Long-legged buzzard *Buteo rufinis*	1,100-5,100$_a$*	6-25	~	
Steppe eagle *Aquila rapax*	?	<1-5	—$_B$	
Lesser kestrel *Falco naumanni*	13-15,000$_a$	26-50	—	1
Lanner *Falco biarmicus*	200	<1-5	-	
Saker *Falco cherrug*	1,200	6-25	-	3
Red-legged partridge *Alectoris rufa*	>100,000	100	~$_{jBC}$	4
Grey partridge *Perdix perdix*	>1,000,000	51-75	—	2
Quail *Coturnix coturnix*	?	26-50	—	3
Crane *Grus grus*	21-22,000$_a$	26-50	-	3
Demoiselle crane *Anthropoides virgo*	?	6-25	—$_B$	3
Little bustard *Tetrax tetrax*	[58-78,000]$_b$	26-50	—$_D$	1
Great bustard *Otis tarda*	15-19,000	51-75	—	1
Stone curlew *Burhinus oedicnemus*	16-20,000$_a$	26-50	-	3
Collared pratincole *Glareola pratincola*	4,500-5,500$_a$*	6-25	-	3
Black-winged pratincole *Glareola nordmanni*	?	51-75	—$_E$	2
Sociable plover *Chettusia gregaria*	?	6-25	—$_E$	1
Black-bellied sandgrouse *Pterocles orientalis*	5,100-10,000$_c$*	6-25	=$_F$	
Pin-tailed sandgrouse *Pterocles alchata*	5,200-20,000$_c$	6-25	-	3
Barn owl *Tyto alba*	97-100,000$_d$	6-25	—	3
Little owl *Athene noctua*	97-160,000$_e$	26-50	—	3
Short-eared owl *Asio flammeus*	18-30,000$_f$	6-25	=	
Dupont's lark *Chersophilus duponti*	500-1,000	26-50	—$_G$	3
Calandra lark *Melanocorypha calandra*	?	26-50	-	3
White-winged lark *Melanocorypha leucoptera*	?	26-50	—$_G$	2$_i$
Black lark *Melanocorypha yeltoniensis*	?	26-50	?	?
Short-toed lark *Calandrella brachydactyla*	?	6-50	=	
Lesser short-toed lark *Calandrella rufescens*	>10,000$_g$	6-25	?	?
Crested lark *Galerida cristata*	?	26-50	-	3
Thekla lark *Galerida theklae*	>110,000	26-50	=	
Skylark *Alauda arvensis*	>10,000,000	26-50	-	3
Tawny pipit *Anthus campestris*	?	26-50	-	3
Meadow pipit *Anthus pratensis*	>10,000,000	76-99	~	4
Isabelline wheatear *Oenanthe isabellina*	?	6-25	+$_G$	
Wheatear *Oenanthe oenanthe*	900-990,000$_h$	26-50	-	3
Rook *Corvus frugilegus*	?	26-50	~	
Rose-coloured starling *Sturnus roseus*	?	26-50	+	
Corn bunting *Miliaria calandra*	?	51-75	-$_H$	2

Key

* European population probably greater than 10,000 pairs
a European palearctic total excluding Soviet Union
b Figure refers to birds not pairs
c No information on Turkish population sizes
d No information on German population sizes
e Population of Britain, France, Spain, Belgium and Netherlands only
f No information on Norwegian and Soviet population sizes
g No information on Turkish or Soviet population sizes
h Population of Britain, France, Belgium, Netherlands, Sweden and Finland only
i Holds 50-75% of world population in winter
j Only natural range considered

++ Large increase; increasing in most countries
+ Small increase; increasing in 20-50% of countries with mainly no change in the rest
= No change in most countries
- Small decrease; decreasing in 20-50% of countries with mainly no change in the rest
— Large decrease; decreasing in most countries
~ No consistent trend; both increasing and decreasing in parts of the range
? No information

Population trends information is taken from Hildén & Sharrock (1985) except where indicated by a subscript, in which case the reference is as follows:

A ICBP (unpublished data)
B Cramp & Simmons (1980)
C ICONA (1986)
D Schulz (1985)
E Cramp & Simmons (1983)
F Cramp (1985)
G Cramp (1988)
H Marchant et al. (1990)

Plates 1 and 2: Two European Red Data Book grassland birds of high conservation priority: sociable plover (above) and lesser kestrel [photos: Paul Goriup]

criteria. Four levels of conservation concern are defined and prioritised as follows:

1 Species which are globally threatened. These have been previously identified by Collar and Andrew (1988).

2 Species whose global populations are concentrated in Europe and are at risk in significant parts of their range.

Here, "concentrated in Europe" is defined as species with more than 50 per cent of their global breeding or non-breeding population in Europe; "at risk" is defined as declining in at least 20 per cent of countries (*cf.* Hildén and Sharrock 1985), or rare with less than 10,000 breeding pairs or 50,000 non-breeding birds, or confined either when breeding, migrating, roosting or wintering to a small number of sites such that more than 50 per cent of the population are within Important Bird Areas (*cf.* Grimmett and Jones 1989).

3 Species whose global populations are concentrated outside Europe, which are nevertheless characteristic of the region and are in need of conservation measures in significant parts of their range.

This category includes those species which are declining in numbers or range, rare or confined to a small number of sites (as defined above), but with European populations below 50 per cent and above 5 per cent of their global populations.

4 Species whose global populations are concentrated in Europe, which are currently thought to be secure in Europe.

This analysis has drawn mostly on published data collated by Eames and Allport (ICBP unpublished) and Allport (ICBP unpublished) and includes data on all regularly occurring species within mainland Europe, Fennoscandia, Turkey and USSR to the border of the Western Palearctic zone. The assessment of the percentage of the world population within Europe was based upon breeding data where available (i.e. the raptors and the remaining bustards). However, for species covered breeding data are incomplete, and for these, calculations were based on mapped distributions indicated in the *Handbook of the birds of Europe, the Middle East and North Africa* (Cramp 1985, 1988, Cramp and Simmons 1977, 1980, 1983) or *An atlas of the birds of the Western Palearctic* (Harrison 1982). Although this has limitations because it assumes a constant density throughout a species' range it is the only information currently available for the entire area in question. Quantitative data for the assessment of passage and winter populations are too incomplete for analysis and are not included.

For most species the assessment of population trends is based upon the study of Hildén and Sharrock (1985). This study used standard questionnaires which were sent to all European countries except Andorra, Luxembourg and Monaco. For each species a national correspondent was asked to indicate whether the species' range was currently (i) increasing, (ii) decreasing, (iii) not undergoing any known change, (iv) of unknown status. The information from these countries has then been summarised according to seven categories of population change across Europe as indicated in Table 2. For those species not covered by Hildén and Sharrock or where more detailed data are available (as for the little bustard) information on their population trends were obtained from other relevant texts as indicated.

Table 2 indicates which species are of conservation concern, based on their population size, percentage of global population and population trends. Their population trends and conservation-priority ratings are summarised in Table 3. Four species are designated as globally threatened: lesser kestrel, little bustard, great bustard and sociable plover and all of these are declining. In addition, sociable plover is restricted to grassland habitats during the breeding season (Table 1) and its population size is unknown. Clearly this species urgently requires further research to establish its current status and whether there are immediate threats to its population. Of the other

Table 3: Summary of the population trends and conservation priorities of dry grassland birds in Europe

Conservation Priority	Large increase ++	Small increase +	Stable =	Small decline -	Large decline —	No trend ~	Unknown trend ?	Total
1					4			4
2				2	4			6
3				10	5			15
4						2		2
Total				12	13	2		27
Non-priority	1	2	4	1	1	3	1	13
Unknown							2	2
Grand total	1	2	4	13	14	5	3	42

Table 4: The conservation status of dry grassland birds in Europe with respect to current international conservation measures

Species	Conservation Priority Ratings	EEC	Bern	Bonn
Cattle egret *Bubulcus ibis*			*	
White stork *Ciconia ciconia*	2	*	*	II
Hen harrier *Circus cyaneus*		*	*	II
Pallid harrier *Circus macrourus*	?	*	*	II
Montagu's harrier *Circus pygargus*	2	*	*	II
Long-legged buzzard *Buteo rufinus*		*	*	II
Steppe eagle *Aquila rapax*			*	II
Lesser kestrel *Falco naumanni*	1	*	*	II
Lanner *Falco biarmicus*		*	*	II
Saker *Falco cherrug*	3		*	II
Red-legged partridge *Alectoris rufa*	4			
Grey partridge *Perdix perdix*	2			
Quail *Coturnix coturnix*	3			II
Crane *Grus grus*	3	*	*	II
Demoiselle crane *Anthropoides virgo*	3		*	II
Little bustard *Tetrax tetrax*	1	*	*	
Great bustard *Otis tarda*	1	*	*	II
Stone curlew *Burhinus oedicnemus*	3	*	*	II
Collared pratincole *Glareola pratincola*	3	*	*	II
Black-winged pratincole *Glareola nordmanni*	2		*	II
Sociable plover *Chettusia gregaria*	1			II
Black-bellied sandgrouse *Pterocles orientalis*		+	*	
Pin-tailed sandgrouse *Pterocles alchata*	3	*	*	
Barn owl *Tyto alba*	3		*	
Little owl *Athene noctua*	3		*	
Short-eared owl *Asio flammeus*		*	*	
Dupont's lark *Chersophilus duponti*	3	+	*	
Calandra lark *Melanocorypha calandra*	3	*	*	
White-winged lark *Melanocorypha leucoptera*	2		*	
Black lark *Melanocorypha yeltoniensis*	?		*	
Short-toed lark *Calandrella brachydactyla*		*	*	
Lesser short-toed lark *Calandrella rufescens*	?	+	*	
Crested lark *Galerida cristata*	3			
Thekla lark *Galerida theklae*		*	*	
Skylark *Alauda arvensis*	3			
Tawny pipit *Anthus campestris*	3	*	*	
Meadow pipit *Anthus pratensis*	4		*	
Isabelline wheatear *Oenanthe isabellina*			*	
Wheatear *Oenanthe oenanthe*	3		*	
Rook *Corvus frugilegus*				
Rose-coloured starling *Sturnus roseus*			*	
Corn bunting *Miliaria calandra*	2			

Conservation priority ratings are as identified by this study

International Conservation Measures

EEC = EEC Directive on the Conservation of Wild Birds
Bern = Bern Convention on the Conservation of European Wildlife and Natural Habitats
Bonn = Bonn Convention on the Conservation of Migratory Species of Wild Animals

Key
* = Species included in conventions
+ = Species likely to be added to EEC Convention at next amendment
II = Species on Appendix II of Bonn Convention

species, eight are concentrated in Europe (i.e. have more than 50 per cent of their world population in Europe) and of these six are at risk in significant parts of their range and thus have a conservation-priority rating of 2. Most grassland birds are of conservation-priority rating 3, having populations concentrated outside Europe, but with their European populations at risk. Only 13 (31 per cent) of the 42 grassland species are not currently of conservation concern (although two are of unknown status).

Only seven species have stable or increasing populations in Europe and of these none are of conservation concern. Indeed at least 25 (93 per cent) of the 27 identified species of conservation concern are declining. Clearly, appropriate conservation measures should be taken for all species of conservation concern in those countries where their populations are at risk.

Existing international conservation measures

The current status of European dry grassland birds with respect to existing relevant international conservation measures is summarised in Table 4.

The Directive and Resolution of the Council of the European Community (EC) on the Conservation of Wild Birds requires that the Member States shall apply special conservation measures to those species listed in Annex I of the directive. This includes measures concerning their habitat in order to ensure their survival and reproduction in their area of distribution and requires Member States to designate Special Protection Areas (SPAs) for the conservation of these species. However, designation of SPAs has been slow, particularly for grassland habitats. Of those grassland areas within Europe identified as Important Bird Areas (IBAs) (*cf.* Grimmett and Jones 1989) and of particular importance for globally threatened species, very few have any protection (Table 5).

Many species of grassland birds are dispersed and at low densities such that conservation of these species cannot be effectively carried out by the designation of restricted areas for protection. For these species a conservation approach incorporating wide scale habitat conservation is required. Such an approach is met by the requirement for Member States to establish Environmentally Sensitive Areas. However, to date these areas have been established only in Denmark, West Germany and the United Kingdom, the latter including the dry grassland areas of Breckland (IBA designated site No. 170; 52°28'N 00°35'E) and the South Downs. Environmentally Sensitive Areas have also been proposed in France and would include the important dry grassland area of Plaine de la Crau (IBA designated site No. 131; 43°34'N 04°52'E). It has been suggested that few other countries have adopted the ESA scheme because the EC contribution is too low and poorer Member States are unwilling to find the remaining funds, especially when agricultural development grants provide greater EC support (Mathers and Woods 1989).

The Convention on the Conservation of European Wildlife and Natural Habitats, known as the Bern Convention, requires that signatories take appropriate and necessary measures to ensure the conservation of the habitats of wild flora and fauna including those bird species listed in Appendix II

Table 5: (*opposite*) The protection status of some designated grassland Important Bird Areas of particularly high national priority for breeding populations of globally threatened dry grassland species

---

Figures given for each species are breeding pairs, except where indicated as follows:
[ ]   = individuals
M   = Males

Criteria (breeding pairs) for identifying IBA sites as particularly high national priority:
Great bustard: sites with   250+ in Spain
                             50+ in Portugal, Turkey and Soviet Union
                             25+ in Hungary
                             5+ in others
Little bustard: sites with   250+ in Spain
                             50+ in Portugal and France
                             25+ in Italy
                             5+ in others
Lesser kestrel: sites with   25+ in Spain
                             5+ in others

Protection status:
SPA   EC Special Protection Area   R   Ramsar Site
NR   Nature Reserve   NP   National Park
PLA   Protected Landscape Area   pla   Protected landscape area proposed
NH   Non-hunting area   BR   Biosphere Reserve
GR   Game Reserve   BP   Bustard Protection Area
P   Protection proposed

Parentheses indicate that only part of the area is protected

			Population (pairs)			Protection	
Site	IBA No.	Location	Great bustard	Little bustard	Lesser kestrel	SPA	Other protection
Czechoslovakia							
Znojmo area	13	49°00'N 16°12'E	[30]			-	P
Podunaji	17	48°00'N 18°00'E	[30-50]			-	pla
France							
Plaines de Pons-Rouffiac	71	45°35'N 00°32'W		60			No
Plaine de Niort Sud-est	74	46°19'N 00°27'W		100			No
Plaine de Niort Nort-ouest	75	46°23'N 00°32'W		120			No
Plaine de St.Jouin-de-Marnc et d'Assais-les-Jumeaux	76	46°23'N 00°32'W		90			No
Plaine de St.-Jean-de-Sauves	77	47°01'N 00°05'E		70			No
Plaines de Mirebeau et de Neuville-de-Poitou	82	46°40'N 00°15'E		100			No
Plaine de la Crau	131	43°34'N 04°52'E		400-500$_M$	4-6		No
East Germany							
Steckby-Lödderitzer Forst und Zerbster Ackerland	17	51°78'N 12°04'E	20-50			-	NR,BR,BP
Belziger Landschaftswiesen	21	52°12'N 12°40'E	[50]			-	BP
Hungary							
Hortobágy	16	47°37'N 21°05'E	180-200			-	R, BR, NP
Dévaványa	23	40°01'N 20°58'E	20-25			-	PLA
Italy							
Le Murge di Monte Caccia	96	40°53'N 16°24'E			<25		No
Campo d'Ozieri and surrounding plains	112	40°45'N 08°50'E		[50-100]			No
Altopiano di Campeda	114	40°18'N 08°47'E		35M			No
Media Valle del Tirso	119	40°10'N 08°50'E		[50-70]			No
Portugal							
Upper River Douro	6	41°10'N 06°45'W			25+	No	(NH)
Castro Verde plains	27	37°40'N 08°05'W	500	300		No	(NH)
Elvas Plains	28	38°55'N 07°19'W	[100]	200	10	No	GR
Monforte plains	29	39°03'N 07°26'W	30	50		No	(NH)
Spain							
La Limia	10	42°10'N 07°40'W		[750-800]		No	
Tierra de Campos	37	42°00'N 05°00'W	[2,000]	+		No	
Villiafáfila - Embalse del Esla	38	41°50'N 05°40'W	[1,000]	200+		Part	(GR)
Embalse de Alcántara	101	39°45'N 06°30'W		[200+]	150+	No	
Cuatro Lugares	102	39°40'N 06°20'W	[200]	[2,000]		No	
Zorita - Madrigalejo	110	39°10'N 05°40'W	[300+]	[200+]	25+	No	
Trujillo - Torrecillas de la Tiesa	111	39°30'N 05°50'W	[90]	[500+]	110+	No	
Llanos entre Cáceres y Trujillo	112	39°25'N 06°10'W	[c.1,000]	[2,000+]		Part	
Cáceres - Sierra de Fuentes	113	39°25'N 06°20'W		100+		Part	
Malpartida de Cáceres - Arroyo de la Luz	114	39°25'N 06°30'W		[200+]	120+	Part	
Brozas - Membrió	115	39°40'N 06°50'W	[c.1,000]			No	
Aldea del Cano - Casas de Don Antonio	118	39°15'N 06°20'W	[100-150]	[3,000]		No	
El Membrio - Le Albuera	125	38°50'N 06°50'W	[500+]	200+		No	
Granja de Torrehermosa - Llerna	134	38°20'N 05°40'W	[600+]	[2,000+]		No	
La Serena	137	38°50'N 05°30'W	[500+]	[20,000]		No	
Los Monegros	178	41°25'N 00°15'W	[32]		50	No	
Turkey							
Vicinity of Tuz Gölü	32	38°43'N 33°22'E	>10			-	
Vicinity of Bafa Gölü	34	37°31'N 27°27'E			25+	-	
Acigöl	43	37°49'N 29°52'E	10+			-	NH
Soviet Union							
Vicinity of Borisoglebovka	92	51°23'N 42°02'E	200			-	
Yugoslavia							
Jazovo - Mokrin	34	45°50'N 20°15'E	10+			-	
Klisura Crana reka	52	41°20'N 22°00'E			15-20	-	(NH)

(* in Table 4). The effectiveness of this can be best assessed by the extent to which signatories have given protection to Important Bird Areas. It is clear from Table 5 that for dry grassland species this convention has had little appreciable impact to date.

The fundamental objective of the Bonn Convention on the Conservation of Migratory Species of Wild Animals is to protect migratory species in recognition that this requires international cooperation and action. Parties to the Convention that are Range States of the species listed in Appendix I or Appendix II are obliged to provide strict protection for species in Appendix I and conclude Agreements with other Range States for the conservation and management of species in Appendix II. Appendix I species are those in danger of extinction but many which are globally threatened are excluded, including all globally threatened grassland species. For Appendix II species it requires the maintenance of a network of suitable habitats appropriately disposed in relation to the migration routes of these species. Those globally threatened species included in Appendix II should be a priority and would include lesser kestrel. To date though, no ratified Agreements have been made between the participating nations regarding specific protection measures for any of the species listed under Appendix II. The only progress has been a proposal for an Agreement for the white stork *Ciconia ciconia* which may be ratified in 1991.

The location of European populations of globally threatened grassland species

The national populations of lesser kestrel, little bustard and great bustard are presented in Table 6. The importance of the Spanish populations of these species is clear from this. Only Greece and the Soviet Union also support more than 5 per cent of the European population of lesser kestrel. Although the European little bustard population is quite large they are restricted to just five countries, with the vast majority within Spain. European great bustard populations are currently less restricted than the little bustard, but the total population is smaller and again concentrated within Spain. The Soviet Union is however unique in Europe in having sole responsibility for the breeding population of one globally threatened species, the sociable plover.

The distribution of European populations of dry grassland birds of conservation concern

Figure 1 indicates the number of dry grassland species of conservation concern that occur in each country. The importance of Spain, the Soviet Union and Turkey for grassland species are clear and probably reflects to both the extent of grassland and the relatively low intensity agriculture in these countries.

DISCUSSION

It is clear from this review that most of the European dry grassland birds are undergoing widespread population declines and that many of these species are of particular conservation concern. However, this analysis is incomplete in many respects. Firstly, population size estimates are unavailable for many species and several regions. Even where estimates are available these are rarely based on good

Table 6: National populations of globally threatened grassland birds

	Lesser kestrel[a]		Little bustard[b]		Great bustard[b]	
	Population	European %	Population	European %	Population	European %
Austria					100	<1-5
Bulgaria	<100	<1-5				
Czechoslovakia					315	<1-5
France	10	<1-5	7-15,000	6-25		
East Germany					400	<1-5
Greece	2,000	6-25				
Hungary					2,365	6-25
Italy	200-400	<1-5	c. 2,000	<1-5		
Poland					3	<1-5
Portugal	300	<1-5	?	6-25	1,015	<1-5
Romania	120-130	<1-5			300-350	<1-5
Spain	4-5,000	26-50	50-70,000	76-99	12-14,000	50-75
Turkey	500-2,000	<1-5			145-4,000	6-25
Soviet Union	?	c. 26-50	?	6-25	2,980	6-25
Yugoslavia	10	<1-5			30-40	<1-5

Data are taken from Biber (1990) for lesser kestrel and from Goriup (1987) for bustards, except for Spain for which data are from Alonso and Alonso (1991)

a = pairs
b = birds

(Sociable plover is not included; the entire European population is within the USSR) quantitative data and are more often based on educated guesses with little corroboration. Therefore range information has had to be used to assess population distributions for most species. Similarly, for most countries the assessment of population trends is based on expert opinion rather than long term monitoring studies. Furthermore, the questionnaire approach of Hildén and Sharrock (1985) to assessing population trends has many other limitations (as they recognised), including variability in quality of information between species and countries, differences in size of the countries, and differences in the period over which the population trends are considered. Consequently, a conservation-priority rating cannot be determined for some of the grassland species. For those species with ratings, these should be regarded as tentative, until better quantitative data are available. Such data are currently being collected as part of the ICBP Dispersed Species Project and an updated list will be produced as a result of this.

These limitations clearly indicate the need for quantitative surveying of bird populations and long-term monitoring of their trends. Uniform and reliable data on these are crucial to identifying priority species for conservation, establishing the locations of significant populations of these species and the assessment of the success of conservation and protection measures. Furthermore, accurate monitoring combined with the collection of appropriate environmental data can also indicate possible causes of population change.

Despite the limitations of the data outlined above, it is apparent that the conservation measures applied to date have been ineffective regarding most dry grassland species. The observed declines of most species is almost certainly caused by widespread habitat destruction and degradation. Many species are tolerant to low intensity agricultural development (Table 1) whereby grassland is improved or converted to traditional methods of cereal cultivation ("pseudo-steppe") as remain in some parts of the Mediterranean basin (Goriup 1988). However, further intensification leads to rapid impoverishment of the avifauna. Heavy grazing pressure in arid regions quickly degrades the habitat (Goriup and Parr 1981) and in extreme cases leads to desertification, whilst in temperate regions current highly intensive agricultural practices with high fertiliser and pesticide input, autumn sown crops, improved crop varieties and reduced use of rotation in favour of monocultures has resulted in the decline of several farmland species (O'Connor and Shrubb 1986). Grassland for silage and cereal crops are now too tall and dense for successful nesting in spring by stone curlews *Burhinus oedicnemus* and suitable feeding areas of grazed grassland are often no longer in the vicinity of nesting fields (Green 1988). Reduced use of rotations has limited the nesting opportunities of skylarks, consequently causing reduction in successful broods and population decline (Marchant *et al.* 1990, Schläpfer 1988). Increased use of pesticides on cereals has reduced invertebrate food availability for grey partridge chicks and has been at least partly responsible for their decline in intensive cereal growing areas (Rands 1985, Potts 1986).

Most dry grassland species cannot be conserved by the designation and protection of a few specific areas as these species are generally dispersed and occur at low density. Instead, wide scale habitat conservation measures are needed. Especially given the recent accession of Spain and Portugal to the EC and imminent moves towards a market economy in eastern Europe and the Soviet Union; effects which will almost certainly intensify agriculture in the grasslands of these regions. Clearly the subtlety of the effects of agricultural change on grassland bird communities must be studied and taken into account as well as the requirements of the land-users. Land-use policy reforms such as those incorporated in the concept of Environmentally Sensitive Areas may provide such an opportunity.

CONCLUSION

This review has identified 27 species of dry grassland bird as being of conservation concern. Furthermore, the current data strongly suggests that most grassland species are declining. Four globally threatened species are found in dry grassland habitats: lesser kestrel, little bustard, great bustard and sociable plover, all of which are declining. Additionally, the sociable plover is restricted to dry grassland habitats and within Europe only breeds in the Soviet Union. The remaining three species have European populations concentrated in Spain. Spanish, Soviet and Turkish grasslands are also clearly important for the other species of conservation concern as they hold a substantial proportion of these species.

The present analysis is incomplete as data is unavailable for many species and some regions. Further quantitative data on population size and trends is therefore required to reliably establish which species are of conservation concern, where significant populations of these occur and whether implemented conservation measures are successful. However, despite the limitations of this analysis, it is clear that existing international conservation measures have had little effect on grassland birds and that they are threatened throughout Europe. Conservation initiatives are therefore required that not only give protection to key sites where necessary but also address the broader need for agricultural and land use policies that take account of grassland birds.

ACKNOWLEDGEMENTS

I wish to thank Richard Grimmett and Eduardo de Juana for their advice and constructive criticisms of the draft manuscript. Valuable help in the collation of data was also given by Gary Allport and Guy Duke. I am also grateful to Judy Jackson for typing the tables and Craig Robson for the production of Figure 1.

REFERENCES

Alonso, J.C. and Alonso, J.A. (1991). *Parámetros demográficos, selección de hábitat y distribución de la Avutarda (*Otis tarda*).* Icona, Madrid.

Biber, J-P. (1990). *Action plan for the conservation of western lesser kestrel* Falco naumanni *populations*. ICBP Study Report No. 41. International Council for Bird Preservation, Cambridge.

Collar, N.J. and Andrew, P. (1988). *Birds to Watch: the ICBP World Check-list of Threatened Birds*. ICBP, Cambridge, UK.

Cramp, S. (Ed.) (1985). *Handbook of the birds of Europe the Middle East and North Africa: the birds of the Western Palearctic Vol. IV*. Oxford University Press, Oxford.

Cramp, S. (Ed.) (1988). *Handbook of the birds of Europe the Middle East and North Africa: the birds of the Western Palearctic Vol. V*. Oxford University Press, Oxford.

Cramp, S. and Simmons, K.E.L. (Eds.) (1977). *Handbook of the birds of Europe the Middle East and North Africa: the birds of the Western Palearctic Vol. I*. Oxford University Press, Oxford.

Cramp, S. and Simmons, K.E.L. (Eds.) (1980). *Handbook of the birds of Europe the Middle East and North Africa: the birds of the Western Palearctic Vol. II*. Oxford University Press, Oxford.

Cramp, S. and Simmons, K.E.L. (Eds.) (1983). *Handbook of the birds of Europe the Middle East and North Africa: the birds of the Western Palearctic Vol. III*. Oxford University Press, Oxford.

Goriup, P.D. (1987). Some notes on the status and management of bustards with special reference to the European species. *In:* Farago, S. (Ed.), *The great bustard (*Otis tarda*), nature conservancy and breeding of the protected species*. Proceedings of the Symposium in Budapest on June 2nd 1987, pp. 7-24. International Council for Game and Wildlife Conservation, Budapest.

Goriup, P.D. (1988). The avifauna and conservation of steppic habitats in Western Europe, North Africa, and the Middle East. *In:* Goriup, P.D. (Ed.) *Ecology and Conservation of Grassland Birds*, 145-157, ICBP, Cambridge.

Goriup, P.D. and Batten, L. (1990). The conservation of steppic birds - a European perspective. *Oryx* 24: 215-223.

Goriup, P.D. and Parr, D. (1981). Report on a survey of bustards in Turkey, 1981. *ICBP Study Report* No. 1. ICBP, Cambridge.

Green, R.E. 1988. Stone-curlew conservation. *RSPB Conservation Review* 2: 30-33.

Grimmett, R.F.A. and Jones, T.A. (1989). *Important birds areas in Europe*. ICBP, Cambridge.

Harrison, C. (1982). *An atlas of the birds of the Western Palearctic*. Collins, London.

Hildén, O. and Sharrock, J.T.R. (1985). A summary of recent avian range changes in Europe. *Acta XVIII Congr. Int. Ornith., Moscow 1982*: 716-736.

ICONA (1986). *Lista roja de los vertebrados de España*. Instituto para la conservación de la Naturaleza, Madrid.

de Juana, E., Santos, T., Suarez, F. and Telleria, J.L. (1988). Status and conservation of steppe birds in Spain. *In:* Goriup, P.D. (Ed.) *Ecology and Conservation of Grassland Birds*, 113-123, ICBP, Cambridge.

Marchant, J.H., Hudson, R., Carter, S.P. and Whittington, P. (1990). *Population trends in British breeding birds*. BTO, Tring.

Mathers, M. and Woods, A. (1989). Making the most of Environmentally Sensitive Areas. *RSPB Conservation Review* 3: 50-55.

Petretti, F. (1988). An inventory of steppe habitats in Southern Italy. *In:* Goriup, P.D. (Ed.) *Ecology and Conservation of Grassland Birds* 125-143, ICBP, Cambridge.

Potts, G.R. (1986). *The partridge*. Collins, London.

Rands, M.R.W. (1985). Pesticide use on cereals and the survival of grey partridge chicks: a field experiment. *J. Appl. Ecol.* 22: 49-54.

Rodriguez, J.M. and de Juana, E. (1991). Land-use changes and the conservation of dry grassland birds in Spain: a case study of Almería province. *Dry grassland birds in Europe. Proceedings. of the Seminar held in the UK in April 1991*.

Schläpfer, A. (1988). Populationsökologie der Feldlerche *Alauda arvensis* in der intensiv genutzten Agrarlandschaft. *Orn. Beob.* 85: 309-371.

Wolkinger, F. and Plank, S. (1981). *Dry grasslands of Europe*. Nature and Environment Series, No. 21. European Committee for the Conservation of Nature and Natural Resources, Strasbourg.

LAND-USE CHANGES AND THE CONSERVATION OF DRY GRASSLAND BIRDS IN SPAIN: A CASE STUDY OF ALMERÍA PROVINCE

Juan Manrique Rodríguez[1]
Eduardo de Juana[2]

[1] Instituto de Bachillerato Celia Viñas, 04004 Almería, Spain
[2] Depto. Biología Animal I, Facultad de Biología planta 9, Universidad Complutense, 28040 Madrid, Spain

ABSTRACT

During the last few decades profound changes have taken place in the traditional ways of land-use in Almería, a province of southeast Spain renowned for its semi-arid climate, steppe-like landscapes and associated avifauna. These changes have generally resulted in population declines of most dry grassland bird species. The data gathered during the compilation of an ornithological atlas of the province allowed us to identify the most important areas for the conservation of these bird species, and relate bird population changes to the land-use transformations that have recently occurred. The replacement of low intensity cereal dry farming by almond tree plantations and/or irrigated cultivations, seems to be the single most detrimental factor, other possible threats being the abandonment of cereal fields, the regeneration of scrub because of insufficient grazing, and locally, reafforestation, urbanization and extraction of sand. The five top-priority areas identified cover some 50,000 ha (5.7 per cent of the provincial surface); only two of them (4,000 ha in all) are legally protected. Protection status should be extended to the other areas and extensive cereal farming should be encouraged, possibly through the designation of the priority steppe zones as EC Environmentally Sensitive Areas.

INTRODUCTION

The Iberian Peninsula has a special importance within Europe for its dry grassland or steppe-like habitats and associated bird fauna (de Juana *et al.* 1988). This fact can be attributed to a fairly large size, a dry Mediterranean type of climate and various other ecological and historical reasons that have determined the predominance of low intensity traditional ways of land use (del Cañizo 1960). However, present social and economic factors such as the accession of Spain and Portugal to the EC (Goriup 1985, Baldock and Long 1987, Giró 1989) are leading to a rapid intensification of the agriculture that could pose a real threat to the conservation of several species, i.e. Montagu's harrier *Circus pygargus*, great bustard *Otis tarda*, little bustard *Tetrax tetrax*, black-bellied sandgrouse *Pterocles orientalis*, pin-tailed sandgrouse *P. alchata*, Dupont's lark *Chersophilus duponti*, and lesser short-toed lark *Calandrella rufescens* most of them already included in the Spanish "Red List" (ICONA 1986; see also Parslow and Everett 1981, Goriup 1988, and Grimmett and Jones 1989). The conservation of these species, as demanded by national and international legislation, will probably depend on the adoption of environmentally-orientated agricultural policies, but these in turn need a sound ornithological and ecological knowledge that only in recent years has started to develop (Ena 1987, Tellería *et al.* 1988, de Juana 1989).

In this report we will contribute some information regarding the conservation of steppe-like habitats and their birds in the province of Almería, where one of us (JMR) has carried out a four-year study on the distribution of birds. This province has a special relevance in this context because of its geographical position in the arid southeast tip of Iberia, the driest area in western Europe, holding landscapes very similar to those typical of the neighbouring countries of Morocco and Algeria (Capel 1987, Alcaraz and Peinado 1987). Two of the five main assemblages of the Spanish shrub-steppes (de Juana *et al.* 1988) are largely included within Almería province. We will try to: (i) identify the areas of importance to the conservation of steppe birds; (ii) describe the changes in land-use that are presently taking place; (iii) relate these changes to recent trends in the distribution and numbers of the different species, and (iv) discuss the possible evolution of the steppe-like areas and their birds and establish some priorities for conservation.

DATA SOURCES AND ANALYSIS

There is only one official map available, which dates from 1982, on the land-uses in the province of Almería (MAPA 1982). On the other hand, the statistical yearbooks of the Ministry of Agriculture ("Anuarios de Estadística Agraria", previously "Anuarios Estadísticos de las Producciones Agrícolas", Ministerio de Agricultura, Pesca y Alimentación) give only information for the whole of the province. In order to assess the changes which have occurred in smaller areas we have used the unpublished questionnaires held by the Cámara Provincial Agraria ("Cuestionarios Estadística 1T", Cámaras Agrarias, Ministerio de Agricultura). Although the quality of information from these sources can often appear somewhat dubious, the broad picture that they provide seems to be adequate for our purposes.

There is an almost complete lack of information, published or unpublished, on the past distribution and abundance of the bird species in this province, which has been rarely visited by ornithologists. In most cases we have had to infer likely trends from the comparisons made between different types of habitat or land-use on the basis of present bird densities, although for birds of a certain size, best known to the local people, we can sometimes cautiously utilise comments made by farmers, shepherds or hunters.

The present status of the birds has been determined during the atlas work carried out in four consecutive springs (1984-1987), by a single observer. In each of the atlas units (tetrads of the sheets of the Spanish official map at 1:50,000 scale, averaging 13,643 ha in the province) the same amount of time/effort was spent (16 hours of intense field work in appropriate dates, hours and weather conditions). Within each tetrad a number of samples (point counts) were taken, distributed among the different types of habitat in proportion to their relative territorial representation. The samples containing "important steppe species" (those listed by de Juana *et al.*1989, excepting crested lark *Galerida cristata*, thekla lark *G. theklae*, short-toed lark *Calandrella brachydactyla*, fantailed warbler *Cisticola juncidis*, spectacled warbler *Sylvia conspicillata* and corn bunting *Miliaria calandra*, which are very widely distributed in the province), were used to establish the "steppe areas". Later on, for each of these areas all the available samples were considered, whether or not they contained important steppe species. In this way we have been able to calculate three different relative indices on the distribution and/or abundance of each species: (1) "distribution index" (percentage of samples that include the species); (2) "overall density index" (contacts of the species per hour spent in the area, all the samples taken into account); (3) "encountered density index" (as above, contacts per hour, but taking only samples where the species was recorded). The significance of these indices must be considered in relation to the intrinsic characteristics of the different species (i.e. detectability, gregariousness, diurnal/nocturnal type of activity, etc.), being in general more valuable for territorial passerines. For some scarcer species some additional information on presence/absence, gathered in other ways or in other years, has been used for the completion of the tables.

RESULTS

The 14 "steppe areas" identified (Table 1, Figure 1), show a distribution that almost coincides with the provincial surfaces showing flat or very smooth relief. Their geographical and ecological characteristics (Table 1) are nevertheless quite varied, with altitudes going from sea level to 1,600 m a.s.l. in Sierra de Gádor and 1,900 m in Sierra de Filabres, the average annual temperature ranging from 11°C to 21°C, and the average rainfall from 170 mm to 500 mm per year (Capel 1977). Additional information is given in Table 1 in relation to the types and extent of land-use existing in them; these are given as a percentage of the land that in the 1982 map was included in one or more of four broad categories: "scrub" (includes all types of uncultivated land with non-arboreal vegetation), "arable" (cereal croplands), "trees" (both cultivated and uncultivated arboreal vegetation: olive trees, almond trees, holm oaks, pines etc.), and "orchards" (all other types of cultivation, most of them irrigated). The corresponding ornithological data are presented in Table 2.

Land use changes

Until the advent of mechanised farming practices in the 1950s and 1960s, the main traditional land-uses in most of the province were cereal dry farming combined with extensive sheep and goat farming and, in some areas, orchards or small plots of almond and olive groves. From 1960 to 1988, the areas devoted in Almería province to cereal crops have decreased by one third (from some 67,000 ha to 45,700 ha), whilst the irrigated areas have doubled in size (from 35,700 ha to 69,300 ha) and the almond plantations have increased almost nine times (from 5,600 ha to 49,900 ha). These changes have been far from homogeneous, since the possibilities of agricultural improvements have closely depended on factors such as topography, climate or underground water availablity. With regard to the "steppe areas" considered here, we can distinguish several categories (Figure 2):

(a) In the extreme north of the province, in relatively flat areas with colder (T < 13°C) and more humid climate (P > 400 mm) (Figure 1: TP, CH), cereal cultivation was predominant, with small patches of coarse grassland (*Stipa* spp., *Thymus* spp.) and some bushes or trees (mostly holm oak *Quercus ilex*

Plate 1: "Las Armotaderas", Costa de Gata, Almería [photo: E de Juana]

Plate 2: Crested lark [photo: Paul Goriup]

Table 1: Important areas for the conservation of steppic or dry grassland birds in Almería province, with data on size, average altitude, average annual temperature, average annual precipitation and land use in 1982 (see text)

	size (ha)	mean altitude (m a.s.l.)	T °C	P (mm)	% surface scrub	arable	trees	orchards
Topares (TO)	>25,000	1,050	12.5	400	20	75	5	0
Chirivel (CH)	11,000	1,050	12.8	400	10	60	30	0
Albox - Oria - Taberno (AT)	5,000	600	18.6	325	30	15	35	20
Huercal Overa - Pulpi (HP)	12,000	200	16.5	250	30	20	30	20
Sierra de Filabres (FI)	1,000	1,900	11.0	450	80	10	10	0
Cuevas - Vera (CV)	6,000	100	21.3	200	40	30	10	20
Gergal (GE)	3,500	750	14.5	250	50	30	20	0
Tabernas - Sorbas (TS)	12,000	500	17.3	240	30	40	10	15
Sierra de Gádor (GA)	800	1,600	13.0	500	100	0	0	0
Dalías Interior (DI)	14,000	90	18.0	300	40	10	0	50
Dalías Litoral (DL)	4,000	30	18.3	270	90	0	0	10
Níjar - Carboneras (NC)	15,000	200	18.5	200	40	30	10	20
Níjar Interior (NI)	5,000	150	18.5	190	80	10	0	10
Níjar Litoral (NL)	2,000	30	19.0	170	90	0	0	10

rotundifolia) in hillocks or spots with poorer soils. The situation has remained fairly stable, with just a slight increase in the cultivated areas made possible by the use of tractors. Nevertheless, in CH a milder climate and the use of new varieties of almond that flower later in the spring, combined with better prices for the product since the 1970s, have led to a significant expansion, still continuing, of almond plantations over former cereal fields.

(b) In a second belt to the south, on more hilly, warm and dry terrain (areas AT, HP, CV, TS and GE), the proportion of land under cultivation was smaller, with cereal crops only on small flat areas with comparatively deep soil; orchards and almond groves occurred in the bottom of little valleys, forming ribbons along seasonal creeks (*ramblas*), and the rest of the land was covered by scrub. The changes here have been very pronounced, the cereal being almost completely replaced by almond plantations in some of the areas (AT, GE, most of TS), or by irrigated orchards (fruits, vegetables) in others (HP, CV).

(c) In the very dry (P < 300 mm) and warm (T > 18°C) parts of the pediment, not far from the coast (DI, NI, NC), extensive cereal cultivations coexisted with large, often hilly areas of scrub used for raising sheep and goats, and gathering esparto grass (*Stipa tenacissima*). In some of the areas, especially DI and western parts of NC, an impressive development of modern irrigated crops under plastic has taken place since the 1960s (Hernández Porcel 1987, García-Dory 1991). This development has completely supplanted the former cereal fields and large expanses of the scrub. Where irrigation has not been possible, cereal cultivation has been abandoned, given the very low output, allowing expansion of rough grassland and scrub, although a slight increase in the area of almonds has been detected in NC.

(d) In addition to the above, there are some small areas in the high mountain plateaux (FI, GA), or in the littoral fringe (DL, NL), which are mostly unsuitable for cultivation due to its rocky or sandy soil, that have been used almost exclusively for grazing. In the mountain areas there were recent massive reafforestations with pines at GA, as well as with sisal *Agave* at NL during the 1950s.

The effects on the bird fauna

We will now briefly review the status and trends of the different bird species in the light of the land-use changes that have taken place. Maps on their present distribution are available in Pleguezuelos and Manrique (1987).

The Montagu's harrier is now extremely scarce and local, breeding with certainty only in DL and NL (and not every year in the latter). There is no data about its former status.

The little bustard has vanished from the more transformed areas and is decreasing elsewhere. It needs wide expanses of suitable habitat, which comprises various types of grassland, including shrub-steppe of *Stipa tenacissima*. In TO and NC, where most land is presently under cultivation, it is local, though common in optimal habitat. In DL and NI it attains a wide distribution but at low densities.

The stone curlew *Burhinus oedicnemus* seems quite adaptable, being still present in most of the study area apart from the mountains. However, it is also declining and its low densities at CH, AT and TS could point to local future extinctions.

The collared pratincole *Glareola pratincola* breeds only in the littoral areas, in very low numbers. During recent years there has been at DL a colony of 25-30 pairs, and a few pairs more have also bred on several occasions in NL.

The black-bellied sandgrouse, like the little bustard, has become extinct in the more severely transformed areas (CH, DI, CV?), and there is some concern about the possibility of it surviving at HP, TO, and NC, where in addition to agricultural changes it faces illegal shooting at water holes. As its detectability is

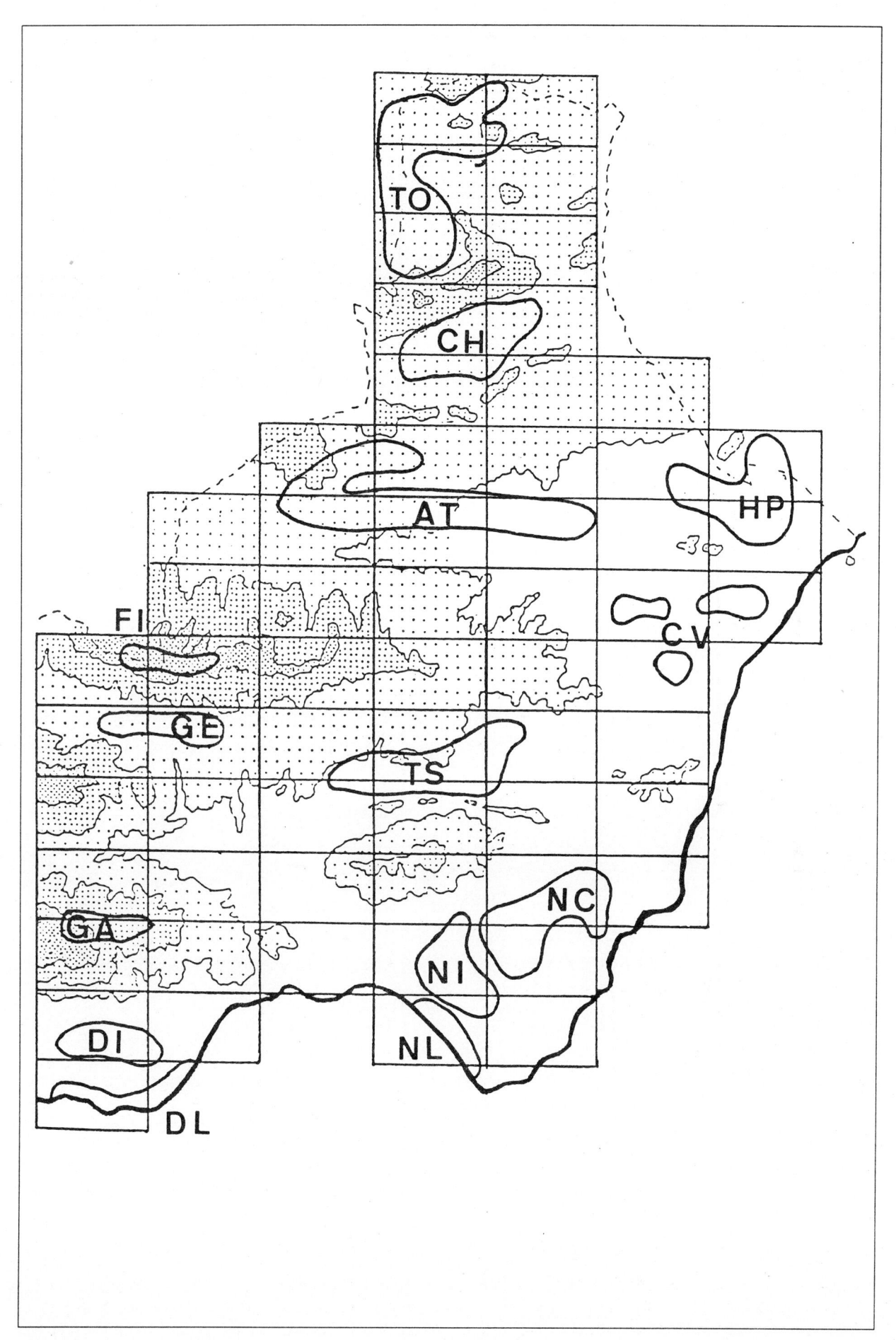

Figure 1: Important areas for the conservation of dry grassland birds in Almería province, Spain (for key, see Table 1 opposite)

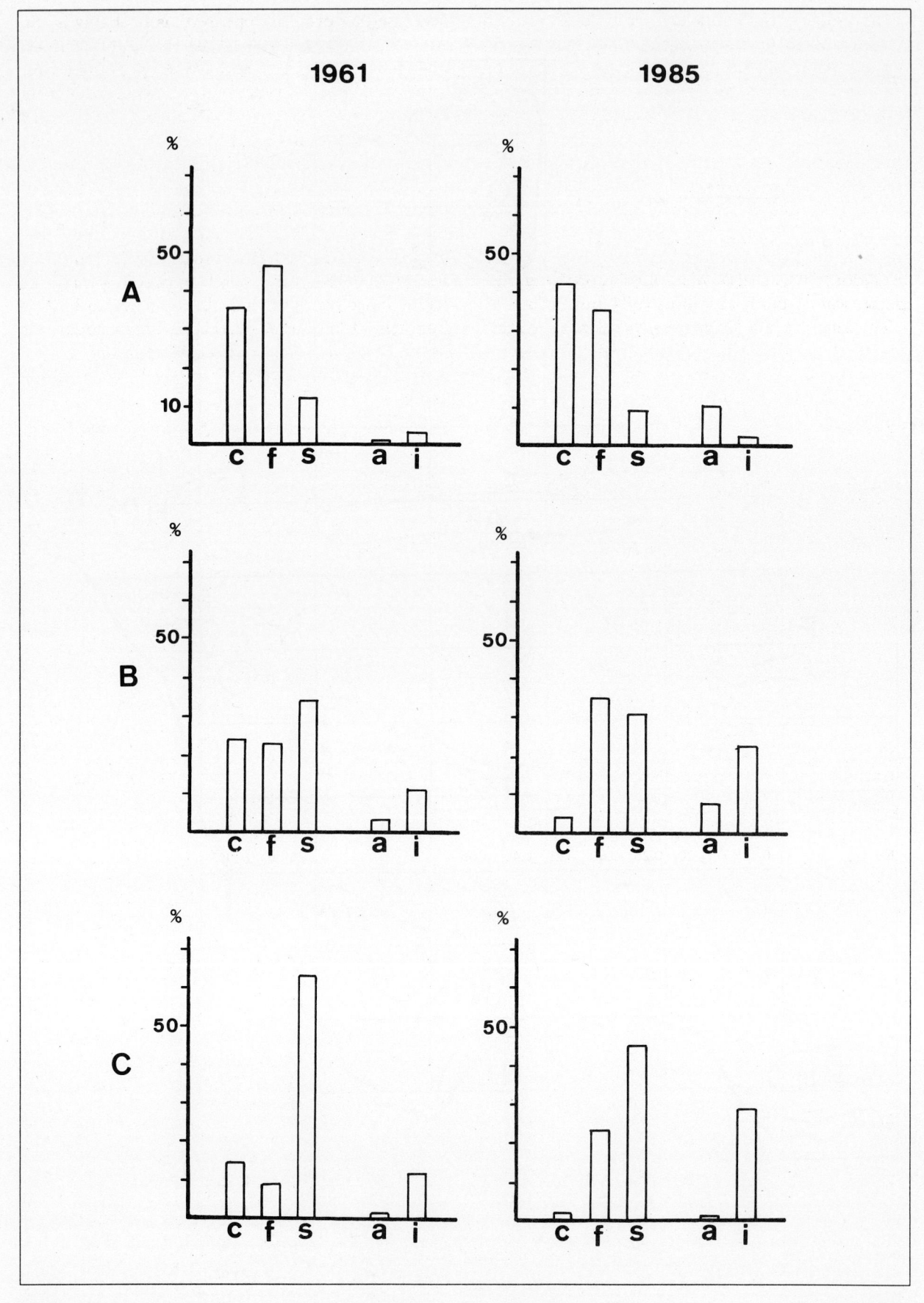

Figure 2: Land-use changes in Almeria between 1961 and 1985, given as percentages of land included in any of the following categories: cereal crops (c), fallow (f), scrub (s), almond plantations (a) and irrigated cultivations (i). Three levels are recognised in the province: (A) North (municipalities of Topares and Chirivel); (B) Central (Albox, Cuevas de Almanzora, Ver, Gérgal, Tabernas and Sorbas); (C) South (Dalías, Almería and Níjar). [Unpublished information obtained from the Cámara Agraria Provincial de Almería]

low and some of our contacts refer to flocks, the real densities could be much higher than our figures suggest. There are no data on the presence in the study area of the pin-tailed sandgrouse, although it seems probable it was once present in TO.

Dupont's lark is extremely local in Almería, where until very recently was only known from NI and NC, breeding on flat *Thymus* and *Stipa* shrub-steppe, avoiding croplands and showing a very patchy distribution. However, in 1990 a new breeding population of some 50 pairs was discovered in Sierra de Gádor (GA), at 1,600 m a.s.l., in a spot where the species was certainly absent in 1984 and 1985.

The calandra lark *Melanocorypha calandra* appears clearly dependent on cereal cultivation. As a consequence, its densities are high in TO (75 per cent of which is arable) and CH (60 per cent arable), while in the rest of the province it has markedly decreased or even disappeared, as probably is the case at AT, HP, CV and GE. In the lowlands there now remain only tiny and scattered populations, some still linked to cereal fields (TS, NC), but others located in marginal grassland habitat.

The short-toed lark is abundant and widely distributed, apparently not much affected by the recent land-use changes. It relies on small patches of land showing a high proportion of bare ground. The lesser short-toed lark shows a clear preference for littoral areas, as well as for steppes with very low, scattered bushes, and much exposed ground. It seems more sensitive than its congener to habitat fragmentation and change; this possibly accounts for its present low numbers at AT, HP and CV.

The Thekla lark is the most widely distributed bird in the province, showing high densities everywhere, both on flat and in hilly terrain. The crested lark, also

Table 2: Ornithological data relative to the important bird areas for the conservation of steppic birds in Almería

	Montagu's harrier D O E	Little bustard D O E	Stone curlew D O E	Collared pratincole D O E	Black-bellied sandgrouse D O E	Tawny pipit D O E
TO	+ + +	8 0.06 0.77	40 0.85 2.09		20 0.18 0.67	24 0.05 1.7
CH		extinct	8 0.15 1.14		extinct	
AT		extinct?	5 0.17 2.67			
HP		extinct	23 0.26 0.88	6 0.04 0.40	6 0.04 1.25	
FI						57 2.8 3.7
CV		+ + +	+ + +		extinct?	
GE		extinct			40 0.50 1.36	
TS		extinct?	11 0.11 0.82		6 0.11 2.67	
GA						72 2.5 3.2
DI		extinct	25 0.27 1.42		extinct	
DL	+ + +	28 0.21 0.71	21 0.21 0.76	+ + +		
NC		8 0.12 0.48	26 0.35 1.45	4 0.04 0.57	8 0.08 0.82	
NI	+ + +	20 0.25 0.80	26 0.40 1.05		26 0.40 1.04	
NL	+ + +	+ + +	37 0.85 2.00		12 0.12 0.88	

	Dupont's lark D O E	Calandra lark D O E	Short-toed lark D O E	Lesser Short-toed lark D O E	Thekla lark D O E	Crested Lark D O E
TO		64 8.6 15.0	72 8.1 11.2		84 9.7 10.5	44 3.9 7.5
CH		42 10.4 16.3	83 12.3 13.6		58 3.9 4.7	42 2.9 5.2
AT			77 6.7 9.7	11 0.6 2.7	56 8.1 12.2	56 4.7 7.6
HP			59 3.7 5.5	41 2.0 4.5	47 8.4 16.5	82 9.4 11.2
FI			86 8.9 9.2		57 1.1 1.7	14 0.1 0.6
CV			60 2.9 4.6	40 2.8 7.2	40 4.4 9.0	53 8.1 12.8
GF			80 6.9 8.2		40 4.4 8.1	40 2.7 5.0
TS		28 5.0 15.3	100 12.9 12.9	6 1.3 14.1	22 10.0 27.0	67 63 14.4
GA	+ + +				72 4.3 5.6	14 0.7 6.0
DI		25 0.8 4.7	100 26.0 26.0	25 1.8 13.3	38 6.9 15.6	63 4.5 7.5
DL		29 1.7 3.7	79 13.0 15.3	58 14.2 21.6	57 6.0 9.3	64 3.6 5.1
NC	+ + +	+ + +	72 7.4 10.0	44 7.1 16.0	48 16.0 28.0	44 6.0 12.6
NI	20 0.7 2.2		47 7.1 12.3	33 5.0 10.2	87 34.0 38.0	13 0.7 10.5
NL			50 9.6 14.0	50 6.2 9.0	75 18.0 26.0	38 6.6 12.0

Key
- D distribution index (percentage of samples in the areas that include the species)
- O overall density index (contacts of the species per hour, all the samples in the area taken into account)
- E encountered density index (contacts per hour, only considering samples that contain the species)
- + presence of the species in the area

very common and widespread, seems to replace the former to a certain extent in certain kinds of habitat, including arable land, open almond and olive groves, and the outskirts of villages, but it does not attain as high local densities as the Thekla. For some 12 per cent of the recorded *Galerida* larks, it was not possible to determine the species involved, these individuals not being included in Table 2.

The tawny pipit *Anthus campestris* is present in just three high-level areas (TO, FI, and GA); its distribution seems to be regulated chiefly by climate. It clearly prefers uncultivated land.

Finally, the trumpeter finch *Bucanetes githagineus*, a North African species that colonised the province in the 1960s and is slowly expanding, breeds in eroded foothills denuded of vegetation, and is only to be seen during the winter in the flat steppe areas preferred by other species.

DISCUSSION

The changes that have taken place in Almería during the last few decades have apparently resulted in a general reduction of the population levels of most of the bird species associated with steppic habitats. Some of them, such as the little bustard and the black-bellied sandgrouse, are now quite rare, having disappeared from several areas, although others such as the stone curlew and most of the larks are still abundant and widely distributed. Considering the population trends of the different species in the study area, together with their situation at international level (Collar and Andrew 1988, Parslow and Everett 1981) and national level (Fernández-Cruz and Araújo 1985, ICONA 1986), we conclude that the priority species for conservation should be Montagu's harrier, little bustard, collared pratincole, black-bellied sandgrouse and Dupont's lark.

The single land-use change that can be identified as most detrimental to the conservation of the steppe-like habitats and their birds in Almería is the progressive substitution of low intensity cereal fields by almond plantations (areas CH, AT, HP, GE, TS, and NC), or irrigated crops (AT, HP, CV, TS, DI and NC). In some places the cereal has simply been abandoned resulting in reversion to dry pasture or scrub, but the loss of the feeding opportunities that are provided by arable (see for example Tellería *et al.* 1989 and O'Connor and Shrubb 1986) could also be negative, the former irregular mosaic of fields and pasture surfaces probably being preferable to uniform expanses of shrub-steppe. Although this aspect should receive further research, it seems clear that, generally speaking, low intensity cereal cultivation should be favoured in Almería in order to halt the decline of the steppe bird populations. This could probably be done through the "Environmentally Sensitive Areas" approach (EC Regulation 797/85). It is also necessary to monitor closely the effects of the reform of the Common Agricultural Policy (Taylor and Dixon 1990), which, seeking a substantial reduction of the EC cereal production, will undoubtedly produce a very serious impact on the Spanish agriculture (Pérez-Tabernero 1987). This will especially apply to semi-arid regions such as Almería, where the average provincial output of the non-irrigated cereal crops in 1988 was only 924 kg per ha for wheat and 1,472 kg per ha for barley (MAPA 1990).

Priority areas

To determine the areas that should have high conservation priority, we have considered for each of the "steppe areas" the species richness, the presence/absence of endangered species (those listed above), the proportion of steppe-like habitat (scrub plus arable *versus* trees plus orchards), the relative integrity or fragmentation of this kind of habitat, and the total area involved. Table 3 presents the results of a provisional system to rate these various characteristics, the relative value of each area being readily apparent. Only a few of them seem to have an outstanding significance, these areas mostly coinciding with those included in the inventory of Important Bird Areas (IBA's) made by SEO for ICBP (Grimmett and Jones 1989).

The area of Topares (TO), in the north of the province, represents approximately one third of IBA No. 223 "Topares-El Moral-Puebla de Don Fadrique", which continues over the neighbouring provinces of Granada and Murcia. It is large (80,000 ha), still supports excellent populations of stone curlew and most larks (especially calandra lark), and smaller numbers, probably endangered, of little bustard and black-bellied sandgrouse. Land use seems fairly stable, but the remaining places with scrub or *dehesa* (scattered holm oaks) should be preserved. Poaching constitutes a local problem. The area does not have legal protection.

The Campo de Dalías Litoral (DL), in the southwest, is the remnant of a formerly wide steppic area of which the inland portions have been greatly altered by the development of irrigated crops under plastic. It mostly coincides with IBA No. 230 "Punta Entinas-Punta del Sabinar-Salinas de Cerrillos", that is partly protected by the Paraje Natural Punta Entinas-Sabinar (1,960 ha, and an EC-designated Special Protection Area, SPA). It harbours small populations of Montagu's harrier, little bustard, collared pratincole and calandra lark, as well as very high densities of lesser short-toed lark. Important local threats are sand extraction for irrigated crops elsewhere, and increasing urbanization in resort areas.

The other three priority areas (NC, NI and NL) are located within the Campo de Níjar in the southeast of the province, together covering some 22,000 ha. This area holds important breeding populations of most steppe bird species, including the regionally rare Dupont's lark and, during winter, important

Table 3: Conservation priority of the different areas following a scoring system that considers:

a) diversity (1 point if 6 to 8 different steppic bird species in the area; 2 points if more);
b) presence of priority species (1 point each: B, little bustard; H, Montagu's harrier; L, Dupont's lark; P, collared pratincole; S, black-bellied sandgrouse);
c) percentage of suitable habitat (1 point if up to 70%-80% of the land is made of scrub plus arable, 2 points if up to 90%-100%);
d) trends (recent evolution of land-uses: minus 1 point if negative, minus 2 points if very negative);
e) homogeneity of the suitable habitat (1 point if relatively continuous);
f) area (1 point if total extension of the area above 4,000 ha, 2 points if above 10,000 ha).

	Diversity	Priority species	suitable habitat	trends	homogeneity	area	TOTAL
Topares (TO)	2	3 (H, B, S)	2	0	1	2	10
Chirivel (CH)	0	0	1	-1	0	2	2
Albox - Oria - Taberno (AT)	1	0	0	-2	0	1	0
Huercal Overa - Pulpi (HP)	1	2 (P, S)	0	-2	0	2	3
Sierra de Filabres (FI)	0	0	2	0	1	0	3
Cuevas - Vera (CV)	1	0	1	-2	0	1	1
Gergal (GE)	0	1 (S)	1	-2	0	0	0
Tabernas - Sorbas (TS)	1	1 (S)	1	-1	0	2	4
Sierra de Gádor (GA)	0	1 (L)	2	-2	1	0	2
Dalías Interior (DI)	1	0	0	-2	0	2	1
Dalías Litoral (DL)	2	3 (H, B, P)	2	0	1	1	9
Níjar - Carboneras (NC)	2	4 (B, P, S, L)	1	-1	0	2	8
Níjar Interior (NI)	2	4 (H, B, S, L)	2	0	1	1	10
Níjar Litoral (NL)	1	3 (H, B, S)	2	0	1	0	7

concentrations of trumpeter finch. The smaller coastal area, Campo de Níjar Litoral (NL), is included within IBA No. 227 "Salinas de Cabo de Gata-Estepa Litoral" (3,000 ha), which in turn forms part of a Parque Natural de Cabo de Gata (26,000 ha, SPA). In addition, since 1989 part of it has been managed as an ornithological reserve by the Agencia de Medio Ambiente de Andalucía and the Sociedad Española de Ornitología (Las Amoladeras, 900 ha). Campo de Níjar Interior (NI) and Campo de Níjar-Carboneras (NC), are both included within IBA No. 228 "Sierra Alhamilla-Campo de Níjar" (50,000 ha, that also includes a small mountain range). They do not have legal protection status. There have been agricultural transformations in them, but the poor availability of water seems to prevent further changes.

In all, the priority areas cover some 51,000 ha out of 116,000 ha of "steppe areas" (respectively 5.7 per cent and 13.1 per cent of the provincial surface). Only some 4,000 ha of the priority areas are currently protected (some 2,000 ha in DL and 2,000 ha in NL). Some sort of legal protection should be afforded to the rest of the priority areas.

Potential threats

Although agricultural transformation is the immediate major threat to the conservation of steppe birds in Almería, there are some other factors that, as we have seen, affect the integrity of certain biotopes (such as urbanization or sand extraction). There are some other potential threats that could in the future become acute problems and should therefore be closely monitored. Two of them deserve our special attention: scrub regeneration and reafforestation.

Scrub regeneration could happen if extensive sheep and goat farming declines, although we still lack information about what grazing intensity is needed in order to maintain more rich and diverse steppe bird communities. The large-scale planting of *Atriplex* bushes (*A. nummularia* and *A. halimus*), as a source of food for sheep, has been initiated at IBA No. 223 by the regional government of Murcia region (García Pérez and Fuentes Blanc (1990); the impact on birds and the future development of this kind of cultivation should be closely monitored. Reafforestation is already a threat in the high-level areas (GA and FI), but these areas are small and certainly of secondary importance; in other areas of the province the semi-arid climate strictly limits the opportunities for wood production. However, some reafforestation schemes could be developed as measures to oppose the "desertification" problem (ICONA 1982, MOPU 1989), which is presently a political priority since 70 per cent of the Almería province is said to be affected by erosion. In this sense it should be stressed that the "steppe areas" we have recognized mostly coincide with flat areas where the risk of erosion is minimal ("stable" areas in the official map of the Instituto Geológico y Minero de España, IGME 1982).

REFERENCES

Alcaraz, F. and Peinado, M. (1987). El Sudeste Ibérico Semiárido. In: Peinado, M. and Rivas Martínas, S. (Eds.) *La vegetacíon de España*, pp. 257-281. Universidad de Alcalá de Henares.

Baldock, D. and Long, T. (1987). *The Mediterranean environment under pressure: the influence of the CAP on Spain and Portugal and the IMPs in France, Greece and Italy*. Institute for Environmental Policy, London, Paris and Bonn.

Capel, J.J. (1977). *El clima de la provincia de Almería*. Caja de Ahorros de Almería, Almería.

Capel, J.J. (1987). Distribución estacional de las precipitaciones en el continente europeo. *Paralelo 37°* **10**: 29-40. Colegio Universitario, Almería.

Collar, N.J. and Andrew, P. (1988). *Birds to Watch. The ICBP World Check-list of Threatened Birds*. ICBP, Cambridge.

de Juana, E., Santos, T., Suárez, F. and Tellería, J.L. (1988). Status and conservation of steppe birds and their habitats in Spain. *In*: Goriup, P. (Ed.) *Ecology and Conservation of Grassland Birds*, pp. 113-123. ICBP, Cambridge.

de Juana, E. (1989). Las aves esteparias en españa. *In: Seminario sobre Zonas Aridas en España*, pp. 199-221. Real Academia de Ciencias Exactas, Físicas y Naturales, Madrid.

del Cañizo, J. and colls. (1960). *Geografía Agrícola de España*. Madrid.

Ena, V. (Ed.) (1987). *I congreso Internacional de Aves Esteparias*. Junta de Castilla y León, León.

Fernández-Cruz, M. and Araújo, J. (Eds.) (1985). *Situación de la Avifauna de la Península Ibérica, Baleares y Macaronesia*. CODA/SEO, Madrid.

García-Dory, M.A. (1991). Agricultura intensiva y explotación de los recursos naturales en el Campo de Dalías. *Quercus* **59**: 43-45.

García Pérez, F. and Fuentes Blanc, E. (1990). *Cultivo y aprovechamiento de arbustos forrajeros*. Consejería de Agricultura, Ganadería y Pesca, Región de Murcia.

Giró, F. (1989). Agricultura y conservación de la Naturaleza. *Quercus* **37**: 39-41.

Goriup, P. (1985). The rain in Spain... *Birds*, winter 1985: 15-20. RSPB, Sandy.

Goriup, P. (1988). The avifauna and conservation of steppic habitats in Western Europe, North Africa and the Middle East. *In*: Goriup, P. (Ed.) *Ecology and Conservation of Grassland Birds*, pp. 145-157. ICBP, Cambridge.

Grimmett, R.F.A. and Jones, T.A. (1989). *Important Bird Areas in Europe*. ICBP, Cambridge.

Hernández Porcel, M.C. (1987). La agricultura intensiva del Campo de Dalías. *Paralelo 37°* **10**: 133-140. Colegio Universitario, Almería.

ICONA (1982). *Paisajes erosivos en el Sureste español. Proyecto LUCDEME*. Monografía No. 26, Instituto para la Conservación de la Naturaleza, Madrid.

ICONA (1986). *Lista Roja de los Vertebrados de España*. Instituto para la Conservación de la Naturaleza, Madrid.

IGME (1982). *Mapa Geocientífico del Medio Natural escala 1/100.000, provincia de Almería, Tomo 2: procesos, riesgos*. Dirección de Angas Subterraneas y Geotecnia, Instituto Geológico y Minero de España, Madrid.

MAPA (1982). *Mapa de cultivos y aprovechamientos de la provincia de Almería. Escala 1: 200.000*. Dirección General de la Producción Agraria y Excma. Diputación de Almería.

MAPA (1990). *Anuario de Estadística Agraria: año 1988*. Ministerio de Agricultura, Pesca y Alimentación, Madrid.

MOPU (1989). *Degradación de zonas áridas en el entorno mediterráneo*. Dirección General de Medio Ambiente, Ministerio de Obras Públicas y Urbanismo, Madrid.

O'Connor, R.J. and Shrubb, M. (1986). *Farming and birds*. Cambridge University Press, Cambridge.

Parslow, J.L.F. and Everett, M.J. (1981). *Birds in need of special protection in Europe*. Council of Europe, Strasbourg.

Pérez-Tabernero, J.J. (1987). El Tratado de Adhesión: sus posibilidades y limitaciones. *El Campo* **104**: 27-46. Banco de Bilbao, Bilbao.

Pleguezuelos, J.M. and Manrique, J. (1987). Distribución y estatus de las aves esteparias nidificantes en el SE de la Península Ibérica. *In*: Ena, V. (Ed.) *I Congreso Internacional de Aves Esteparias*, pp. 349-358. Consejería de Agricultura, Ganadería y Montes, Junta de Castilla y León, León.

Taylor, J.P. and Dixon, J.B. (1990). *Agriculture and the environment: towards integration*. Royal society for the Protection of Birds, Sandy.

Tellería, J.L., Santos, T., Alvarez, G. and Sáez-Royuela, C. (1988). Avifauna de los campos de cereales del interior de españa. *In*: Bernis, F. (Ed.) *Aves de los medios urbano y agrícola en las mesetas españolas*, pp. 173-319. Sociedad Española de Ornitología, Madrid.

DRY GRASSLANDS BIRDS IN FRANCE: STATUS, DISTRIBUTION AND CONSERVATION MEASURES

Patrick Lecomte and Sylvestre Voisin

GEPANA, 26 rue d'Estienne d'Orves, 92120 Montrouge, France

ABSTRACT

Dry grassland birds are widespread in France due to the great variety in climate, soils and relief. The different topographical characteristics, and the farming methods employed, enabled us to define the different dry grassland bird habitats in our country.

We have considered 28 breeding species and five wintering species; others could also be added to this list. The decline in numbers (sometimes leading to the extinction of a species) of dry grassland birds is mainly due to the development of agricultural practices over the past 30 years, and particularly, over the past 10 years.

Few effective conservation measures have yet been put into practice. Only the Crau region might become a protected area. As for future measures, we can only continue to work on developing more ecologically sound agricultural practices, particularly with regard to agricultural plains. Unless there are radical changes to overall agricultural policies, most of the dry grassland birds populations in France will decline dramatically over the next decade.

INTRODUCTION

France, due to its varied climates, soils and reliefs, offers a wide range of habitats for nesting birds (almost 300 species in all), and also for passage and wintering species.

Dry grassland birds, with their special requirements, are able to find a wide variety of habitats in our country well suited to their needs. Of those species proposed by Goriup and Batten (1990), 28 breeding species occur, and even though the great bustard *Otis tarda* became extinct during the Eighteenth century due to over-hunting, there are five additional species which overwinter in France. Several other species found in dry grassland such as the whinchat *Saxicola rubetra* and stonechat *S. torquata* could also be added to that list.

It should be noted that 18 of the species listed for the purpose of this study are included on the French Red List of endangered species (Beaufort 1983). Moreover, two species – the lesser kestrel *Falco naumanni* and the little bustard *Tetrax tetrax* – feature on the CIPO's 1988 world checklist of threatened birds.

France, with its wealth of resources, has an important role to play in the conservation of these birds. The French population of Montagu's Harrier *Circus pygargus* for example, represents over 30 per cent of the overall European population, and the hen harrier *C. cyaneus* nearly 50 per cent (FIR/UNAO 1982).

We decided to limit ourselves to a detailed study of vulnerable or endangered species to allow us to assess fully the seriousness of the situation of dry grassland birds in France. However, some of these species are not strictly dependent on dry grassland in our country. We have, nevertheless, taken them into account because of their vulnerability both in France and the rest of Europe. We have therefore included the crested lark *Galerida cristata*, a species characteristic of the dry grassland plains in France both on the Mediterranean coast and inland; the wheatear *Oenanthe oenanthe*, widely distributed over a large part of coastline and the lapwing *Vanellus vanellus*. This last species is widely distributed, it is not found in highland areas, nor is it dependent on dry grassland, but the information regarding its biology, breeding and the threats which endanger its future relate closely to the situation for grassland birds in general.

We have limited our research to the French continent, excluding Corsica. Sufficient data were gathered to estimate adequately the general situation of dry grassland birds in all types of different habitat. The distribution of some species is illustrated in Figure 1.

STATUS AND DISTRIBUTION OF SPECIES CONSIDERED BY THE STUDY

White Stork
The white stork *Ciconia ciconia* population has been steadily decreasing since the beginning of the century because of changes in habitat, a high mortality rate from collisions with power-lines and over-hunting in its wintering areas (Goriup and Schulz 1990). It was previously found mainly in Alsace. There are also occasional breeding records for the northeast and on the coast. *Migratory/breeding (highly threatened).*

Hen harrier
The hen harrier is widespread throughout marshy habitats and heathlands of France with the exception of the southeast and the Mediterranean plains. Habitat destruction has forced the species to nest in young conifer plantations. *Migratory/breeding (threatened).*

Montagu's harrier
The Montagu's harrier is mainly found in marshlands and polders. Development and destruction of these habitats has resulted in a considerable population decrease. Sixty per cent of the total population is concentrated within 20 per cent of the country (Yeatman 1976). The Montagu's harrier adapts easily to nesting in intensive crops. However, breeding success there is seriously affected by modern agricultural practices. Destruction of their wintering range in the Sudan and Sahel regions of North Africa is also affecting their numbers (Yeatman 1976). *Migratory/breeding (threatened).*

Lesser kestrel
This small gregarious falcon breeds in increasingly declining numbers (Yeatman 1976, Biber 1990). In 1990, this species could only be found in the Crau. The destruction of its habitat (due to irrigation for arboriculture) poses a serious threat to its future. *Migratory/breeding (highly threatened).*

Crane
The crane *Grus grus* bred in the damp heathlands of southwest France up until the end of the Nineteenth century. Cranes are mainly found in France during migration or as a wintering species (in Lorraine, Champagne and Landes). Since 1985 a pair of birds have been breeding in Normandy, but this recolonisation is already under threat due to agricultural development (Moreau 1990). *Migratory/breeding (at present numbers are stable).*

Table 1: Summary of habitat preferences, status and threats to grassland birds in France

Species	Distribution	Status	Threats	Population
White stork *Ciconia ciconia*	wet meadow, crops	stable, threatened	habitat change, hunting in winter localities	about 100 prs
Hen harrier *Circus cyaneus*	heaths, coniferous plantations, crops	threatened	habitat losses	2,800-3,800 prs
Montagu's harrier *Circus pygargus*	wet meadows, crops	stable threatened	agricultural methods, hunting in winter locs.	3,00-4,000 prs
Lesser kestrel *Falco naumanni*	Mediterranean	highly threatened	habitat change (Crau)	5 prs in 1990
Crane *Grus grus*	wet meadows, crops	stable	drainage, habitat change	1 pr
Little bustard *Tetrax tetrax*	thermophilic continental	highly threatened	agricultural methods	<6,000 calling males
Stone curlew *Burhinus oedicnemus*	thermophilic continental	threatened	agricultural methods	<1,000 prs
Collared pratincole *Glareola pratincola*	Mediterranean halophytic	highly threatened	habitat change, localised	about 20 prs
Black-winged pratincole *G. nordmanni*	Mediterranean halophytic	highly threatened	habitat change	very rare
Thekla lark *Galerida theklae*	Mediterranean	stable	habitatchange, very localised	<100 prs
Pin-tailed Sandgrouse *Pterocles alchata*	Mediterranean	threatened	habitat change (Crau)	170 prs
Calandra lark *Melanocorypha calandra*	Mediterranean	threatened	hunting	<1,000 prs
Short-toed lark *Calandrella brachydactyla*	mainly Mediterranean, coastal	stable, threatened	very localised in extensive plains	<1,000 prs
Tawny pipit *Anthus campestris*	mainly Mediterranean, coastal	stable	habitat change	>10,000 prs
Short-eared owl *Asio flammeus*	heaths, dunes	threatened	habitat change & loss, dispersed distribution	<100 prs
Lapwing *Vanellus vanellus*	meadows, crops	threatened	agricultural methods, habitat change	<20,000 prs
Wheatear *Oenanthe oenanthe*	warm coast, mountains	threatened	habitat change, coastal development	<100,000 prs
Crested lark *Galerida cristata*	extensive plains, large alluvial valleys, coast	threatened	habitat change, coastal development	<100,000 prs

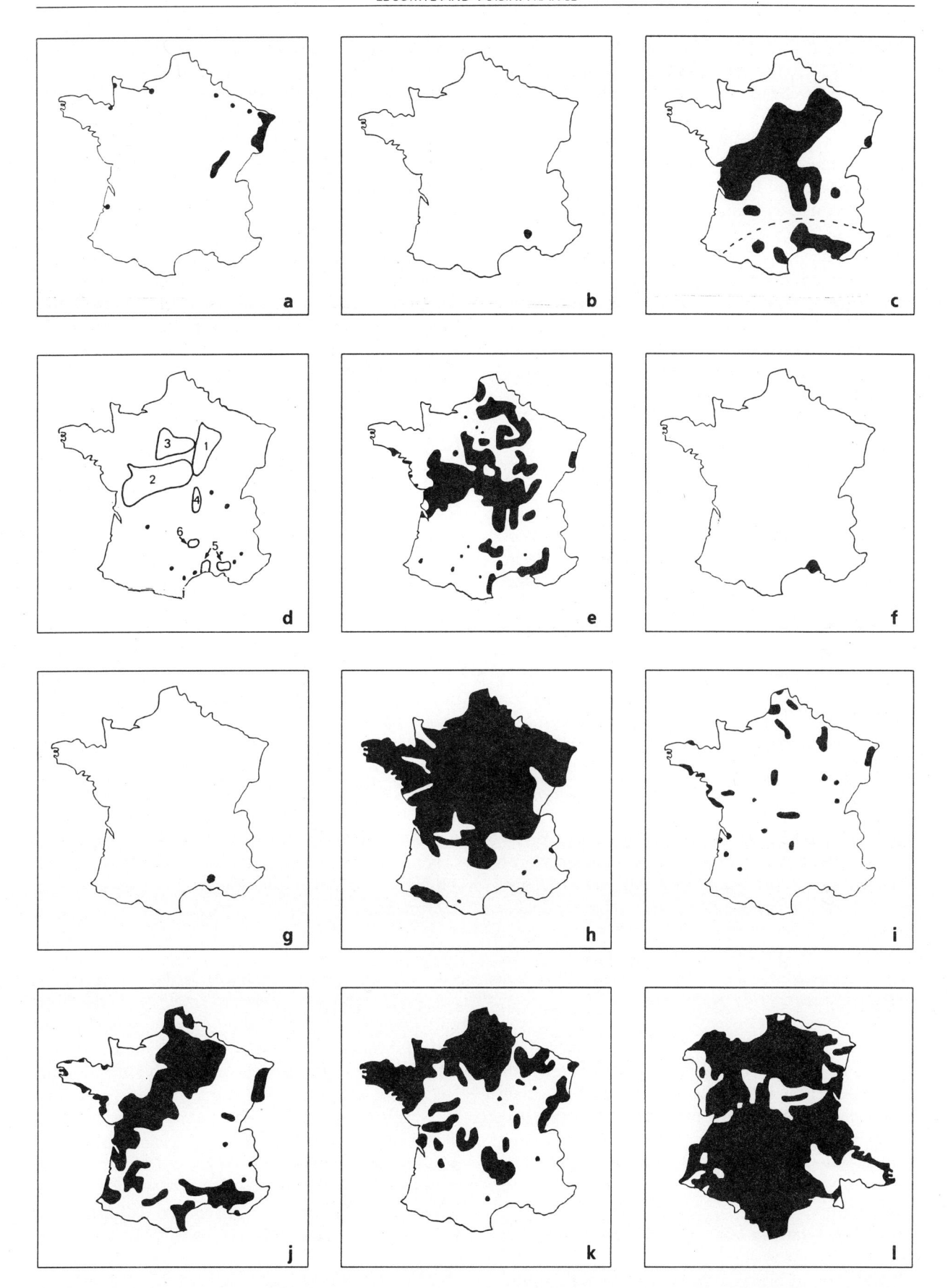

Figure 1: Distribution of some dry grassland bird species in France: (a) white stork (1976); (b) lesser kestrel (1990); (c) little bustard (1976); (d) little bustard (1988): [1 = Champagne; 2 = Champagne berrichonne-Poitou; 3 = Beauce; 4 = Plaine de l'Allier, Limagne; 5 = Languedoc-Provence (Crau); 6 = Causses; large dots = local breeding sites, 1-10 males]; (e) stone curlew (1976); (f) collared pratincole (1976-1990); (g) pin-tailed sandgrouse (1976-1990); (h) grey partridge (1976); (i) short-eared owl (1976); (j) crested lark (1976); (k) meadow pipit (1976); (l) corn bunting (1976)

Little bustard
Found in dry, rocky steppe, although the little bustard has adapted to living in crops. The mechanisation of agriculture, modern cropping and the destruction of grasslands (particularly since 1980) has led to the rapid decline of the species. The little bustard is only commonly found in a 12,000 ha area in the Crau. *Partly migratory/breeding (highly threatened).*

Stone curlew
The stone curlew *Burhinus oedicnemus* is found in steppe areas with scarce vegetation and has probably only survived thanks to an increase in the cultivation of maize and sunflowers. Rarely found in grassland areas, its decline is similar to that of the little bustard. *Partly migratory/breeding (threatened).*

Collared pratincole
The collared pratincole *Glareola pratincola* is a Mediterranean species found in halophytic areas, mainly in the Camargue. The agricultural and industrial development of the Camargue is posing a serious threat to the survival of the species (Walmsley, pers. comm.). *Migratory/breeding (highly threatened).*

Black-winged pratincole
The black-winged pratincole *Glareola nordmanni* is found in Mediterranean saline regions and is occasionally found in the Camargue during migration. The species bred with a collared pratincole once in 1970 (Walmsley 1970, 1976, 1988, 1990). *(Highly threatened).*

Pin-tailed sandgrouse
The pin-tailed sandgrouse *Pterocles alchata* is now only found in steppe in the Crau; it was previously also found on the Mediterranean coastline until agricultural development of its habitat led to a decline in the species. It is likely that this will also threaten its future in the Crau. Its habitat diminished by one third between 1980 and 1989 (Cheylan 1990). *Non-migratory/breeding (highly threatened).*

Short-eared owl
The short-eared owl *Asio flammeus* does not normally nest in dry grassland areas in France. It is usually found in dunes and saline meadows in coastal areas. Its breeding inland is localised and occasional. *Migratory/breeding/overwintering (threatened).*

Calandra lark
The large calandra lark *Melanocorypha calandra* is found in increasingly fewer numbers in the Mediterranean plains. It is under a growing threat from hunting, one of the main reasons for its scarcity. *Non-migratory/breeding (threatened).*

Short-toed lark
The short-toed lark *Calandrella brachydactyla* is particularly common in dry environments. It occurs on the Mediterranean coastline. Populations on the continent disappeared in the Nineteenth century. A few pairs of localised birds remain in intensive cereal plains living in unfarmed areas (e.g. beet storage areas, heaps of stones) (Muselet 1981, 1984). *Migratory/breeding (at present numbers are stable).*

Thekla lark
The thekla lark *Galerida theklae* is native to mountainous scrubland close to the Mediterranean (but not specifically associated with dry grassland). The distribution of this species is concentrated south of Roussillon. *(At present numbers are stable).*

Tawny pipit
The tawny pipit *Anthus campestris* inhabits hot, dry and sparsely vegetated regions. It occurs in areas up to 1,000 m in the south of France and nests in plains elsewhere. It has disappeared from the northern plains of the country, despite the presence of suitable habitats (Yeatman 1976) and its distribution range is now steadily receding towards the south. *Migratory/breeding (at present numbers are stable).*

Crested lark
The species inhabits dry treeless areas and can be found in plains, dunes, chalk downland and large alluvial valleys throughout France, in increasingly declining colonies. *Non-migratory/breeding (at present numbers are stable).*

Wheatear
Found in open stony grassland. It is widely distributed from sea level to the highest alpine meadow. Agricultural development in heathland areas and urban construction/development on coastlines has led to a decline in the species (Yeatman 1976). *(Threatened).*

Lapwing
The lapwing is found in marshland, damp heathland, saline meadows and water meadows. It is the only wader that has adapted to arable farming, but modern agricultural practices have led to a substantial decrease in the population over the last 30 years (Thonnerieux 1987). *(Threatened).*

Status and distribution of dry grassland birds: populations and trends

Using the Atlas of birds breeding in France (Yeatman 1976) and the data we have collected during our study, we have been able to estimate the status of the dry grassland birds we have selected (see Table 1).

Important dry grassland regions

Some regions of France are frequented by several of the above-mentioned species, sometimes in the same habitat. We shall now look at several regions which illustrate the different representative habitats available to dry grassland birds in France (Figure 2).

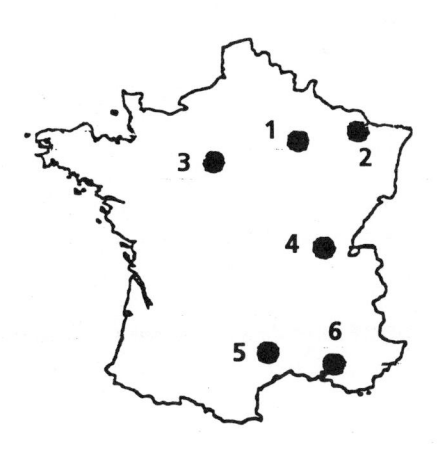

Figure 2: Location of some dry grassland regions discussed in the text: (1) Champagne; (2) Lorraine; (3) Beauce plain; (4) Saone valley plains; (5) Causses region; (6) Crau region

The Crau region
The Crau is a vast steppe region (of 12,000 ha) situated on the old delta of the river Durance with a coarse silty soil. Many species which inhabit this region do not occur elsewhere in France. The number of little bustards in the area are at present stable, and may even be slightly increasing, whereas the species is declining everywhere else. Other birds breeding in the region include stone curlew, tawny pipit, skylark *Alauda arvensis*, calandra lark, short-toed lark, little owl *Athene noctua*, crested lark, red-legged partridge *Alectoris rufa* and Montagu's harrier. Two similar regions in Provence are populated by an identical avifauna (with the exception of the sandgrouse and the lesser kestrel): the Vinon aerodrome (approximately 1,000 ha) and Valensole plateau (approximately several thousand hectares). Since the Eighteenth century, the Crau region has shrunk from 60,000 ha to a mere 12,000 ha and remains under threat from agricultural development (G. Cheylan, pers. comm.)

The Beauce plain
The Beauce plain is a region of cereal production situated on brown acid soils and rendzinas. It has a low rainfall (less than 700 mm per year). The type of intensive agriculture is similar in all of the Paris basin. The species found in this area are of significant interest, but their numbers remain low due to the farming methods. A few pairs of Montagu's and hen harriers are left (FIR/UNAO 1982), and short-toed larks can be found in secluded areas of farming, e.g. beet storage areas, heaps of stones (Muselet 1984): they are highly threatened.

The plains of the valley of the river Saone
Situated on wet soils, these plains are mainly covered by grassland, especially silage grassland. The high rainfall and modern agricultural methods make it particularly difficult for grassland birds to adapt to this region (Broyer *et al.* 1988). The following species are found in this area: little bustard (in drier areas), stone curlew (found in arable farming land) and species that are not dependent on dry grassland such as curlew *Numenius arquata* and lapwing (Broyer *et al.* 1988).

The Causse region
Situated on rendzina soils, the Causse region is a vast area of dry, stony, fragmented grassland. The conditions, rainfall and temperatures of the foothills are very different to that of the dry habitat in the plains. Dry grassland species found in this region have low populations relative to the potential of the habitats.

In the biotopes we have just studied, some species are particularly sensitive to general modifications of the habitat. Such species indicate the degree to which the biotope can support a more or less rich group of species among the ones we are interested in: the little bustard is the first to leave its territories when its habitat has undergone modifications linked with agriculture (disappearance of grasslands, intensification, extreme simplification of crop rotation). Its presence in high densities indicates a high degree of "suitability" of the environment.

The stone curlew and the quail adapt far more easily to changes in their habitats linked with agricultural practices. In the early stages they even take advantage of open fields. Their disappearance indicates that the agricultural plain is becoming extremely "ordinary". It is only necessary to examine the population and distribution of the crested lark, a bird characteristic of dry grassland, to assess the potential of a region. Some species may survive in areas such as aerodromes and military bases. These areas have become the last place of refuge for the little bustard in many regions. Occasionally these areas are found to be islands of rich fauna isolated within the heart of a habitat under destruction. Stone curlews, short-eared owls and the occasional little bustard often occur.

In France, garrigues support species which are dependent on dry grassland in the main part of their distribution area. The remaining populations of these species can be found in garrigues whose conditions best correspond to that of their typical habitat (thekla lark, black-eared wheatear *Oenanthe hispanica*). In France, these species cannot be considered as grassland birds.

The most basic species assemblages can be observed in intensive agricultural plains. They consist of grey partridge (and/or red-legged partridge), corn bunting *Miliaria calandra*, skylark, quail *Coturnix coturnix*, yellow wagtail *Motacilla flava* and stonechat.

In France, many species listed as vulnerable dry grassland birds only inhabit secondary dry grasslands as defined in the project; the hen harrier prefers heaths, the crane prefers wet grasslands and Mediterranean species like the short-toed lark inhabit the garrigue.

HABITAT ANALYSIS OF DRY GRASSLAND BIRDS IN FRANCE

A common factor of all lowland dry grassland birds in France is that their species distribution is not limited to lowland dry grassland, salt marshes and dunes. A study of physiographic features, agricultural activities (arable cropping, viniculture and afforestation) and rainfall (more than or less than 700 mm per year) enable us to obtain an overall view of France's potential as regards salt marsh and dune habitats.

Pedology, climate and relief

Rendzinas and other calcareous soils, which are not very fertile, occupy montane and lowland areas. Maps of annual rainfall were used to locate the driest of these calcareous soils districts.

Despite a climate typical of that of a mountain region (high rainfall with a wide range of temperature), the Causse region (south of the Massif Central), the Jura and Provençal hinterland benefit from favourable conditions with the presence of xerophilic flora and fauna.

Other areas that prove to be suitable habitats, although not situated on rendzinas, are valleys on alluvial soils (e.g. the Plain of Ain) and areas on brown calcareous shallow soils (e.g. the Plain of Limagne). These areas benefit from low rainfalls due to their geographical situation. With the help of this chart and agricultural statistics, we have been able to define lowland dry grassland areas in France for the period 1975 to 1980 (Table 2, Figure 3).

Development and intensification of agricultural practices

From the 1950s onwards, agriculture entered into a period of great change, with increased mechanisation, new storage techniques, increased field size and increased use of artificial fertilisers and herbicides.

Using the agricultural statistics collected, we have attempted to map the overall picture of agricultural practices in 1970, 1979 and 1985 and examined how these different practices have developed during this period (Figure 4).

We noticed the expansion of intensive farming of crops such as maize and wheat, to the detriment of permanent grassland and, in particular, mixed cropping and livestock farming.

Effects of agricultural development on grassland birds

Intensively farmed plains

The environment has become increasingly altered due to the following factors: land improvement, disappearance of hedges and woodlands, monoculture development—sometimes to the extent of almost the whole hectarage of a given region (e.g. the Beauce Plain) — and disappearance of fodder crops such as alfalfa which often offered a final refuge to some species before they died out completely (e.g. the little bustard).

The stone curlew inhabits short vegetation and is commonly found in maize and sunflower crops. When it returns from migration these are exactly the right height for it, whereas other crops are already too well developed. There follows a short breeding period in which it lays and hatches its eggs and rears its young before the height of the crop forces it to move on (Broyer et al. 1988).

The repeated ploughing of soils damages their structure and reduces their capacity to absorb water (De Ploey 1990). Consequently, grassland species which nest at ground level (which is often the case), see their clutches inundated by rain after a downpour.

In the above-mentioned regions, birds are also affected both directly and indirectly by the use of agrochemicals.

The information that we were able to obtain through agricultural and ecological studies shows that the combination of these factors has led to a dramatic decline in bird populations in intensive agricultural plains. At first, the species managed to

Table 2: Types of dry grasslands (on lowland plains) in France defined by the present study

Soils	Rainfall	Agriculture (1976-1980)	Grassland type
typical rendzinas	<700 mm	grasslands	I
rendzinas mixed with other soils	<700 mm	grasslands	I
typical rendzinas	<700 mm	mixed farming (arable/livestock)	I
other soils	<700 mm	grasslands	II
rendzinas mixed with other soils	<700 mm	mixed crops	III
typical rendzinas	>700 mm	mixed crops, grasslands	III
typical rendzinas	<700 mm	intensive agriculture	IV
rendzinas mixed with other soils	<700 mm	intensive agruclture	IV
rendzinas	>700 mm	intensive agriculture	V
rendzinas	<700 mm	woods, maquis/vineyards	VI

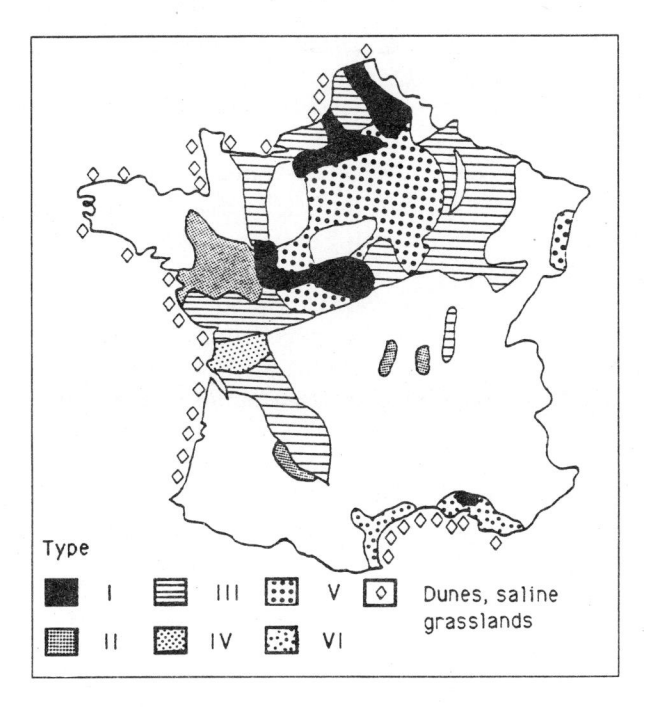

Figure 3: Distribution and types of grasslands in France

adapt to the changes but once the damage to their habitat had reached a certain level, their population rapidly started to decline. It would appear that in the Beauce and other similar areas the decline in numbers began in the early 1980s.

Grasslands
These environments have also been affected by significant changes in agricultural practices (Broyer *et al.* 1988). The use of fertilisers has become more and more frequent. These speed up the growth of the vegetation and bring forward the hay-making season, and as a result, nests are destroyed during harvesting.

Grass is no longer stored as it was between the 1960s and the 1980s. It is now made into silage. This procedure means that harvesting takes place even earlier (Broyer *et al.* 1988) and the number of cuttings is increasing.

Fodder crops are more compact and maize and annual fodder crops have become a particularly common choice in preference to fallow grassland. This is partly due to the common practice of "soil-less" livestock farming (INRA 1989).

CONSERVATION

Current conservation measures and projects
France has signed and ratified several Conventions: for example Ramsar, Bern, Bonn and Lisbon (the last concerning Mediterranean environments). But when we consider the protective measures that have been brought into effect to date (e.g. only one protected site under Ramsar: the Camargue — and only a part of it at that), we end up asking ourselves whether it is really worth pursuing these measures.

For a while now, those involved in gamebird management were concerned that the numbers of gamebirds (e.g. quail and partridge) were diminishing in the intensive agricultural plains and measures have been undertaken to preserve these species, with the creation of protected strips of land, winter feeding and installation of water-holes in summer. These measures have shown positive results for partridge, but have been less effective for quail and have no effect at all on the numbers of little bustard and stone curlew.

Over the past 10 years, the Intervention Fund for Raptors has organised wardening of nests of threatened species (particularly harriers) breeding in arable crop areas. Several different methods of protection exist, including relocation of young birds during harvesting to reconstructed nests in the same or neighbouring unharvested fields and monitoring of the young until they are able to fly. Some nature reserves have also been set up to protect the habitats of some of the birds listed.

Few conservation measures have yet got underway. Those that have concern particularly rich environments such as the Crau, the protection of seriously threatened species at their nesting sites, such as the white stork and the little bustard, or during migration, for example, for the crane. The aim is to protect certain areas from agricultural development which is likely to harm the bird's habitat and to resolve conflicts (e.g. where birds are damaging crops).

These measures, which generally consist of managing the different sites, are often introduced by local, departmental or regional associations which are helped by national associations (WWF France, LPO, FIR), the Ministry of Environment and Regional and General Councils and the EC.

The "Conservatoire National du Littoral" protects dunes and salt marshes that it has bought. The "Conservatoires Regionaux du Patrimoine Naturel" have an important role to play in ensuring that the conservation measures are followed through and in managing both bought and leased sites.

The Ministries of Agriculture and of the Environment have started to look into the more common problems but they still lack expertise in this field.

Aerodromes, whether military or otherwise, and military bases often prove good conservation areas for some dry grassland species without further conservation measures generally being necessary.

Lastly, several organisations such as the CNRS and the INRA have undertaken research following-on from the introduction of the EC Set Aside Policy. Unfortunately, the results will only be available in a couple of years' time.

The Crau Region
The area of the Crau region has diminished from 60,000 ha in the Sixteenth century to 12,000 ha

today. This habitat is particularly rich in endangered species and has been under partial protection for the last few years. There were only 420 ha under protection at the end of 1990:
- 160 ha, including 60 ha of *coussous* (steppe) in a nature reserve created in 1988;
- 160 ha bought by a horticulturist in 1989 and by WWF, CEEP, SEN and CRPNPO in 1990;
- 100 ha bought by a livestock farmer in 1990, funded by the EC and the Ministère de l'Environement.

The listing of another area, representing 100 ha, as a nature reserve should protect the last remaining colony of lesser kestrel. The land bought by a horticulturist was listed as a protected area in August 1990 also to protect the species.

EC Funds awarded in 1990 and the intervention of the FIQV (5.6 million francs) in accordance with Article 19 covering 2,000 ha (800,000 francs) means that it will be possible to protect larger areas of land by selling it to livestock farmers, conservation organisations and the Conservatoire du Littoral. Subsidies of 200 to 400 francs per hectare will be made to landowners who are prepared to undertake conservation management over the next five years. However, no measures have yet been taken to protect the Valensole plateau, which is another exceptional area of Provence (G. Cheylan, pers. comm.).

The Champagne region
This region is inhabited by white storks (rare), and significant numbers of little bustards and cranes during migration or wintering. Several conservation measures have been undertaken by the Conservatoire du Patrimoine Naturel de Champagne-Ardenne:
- Leasing of a site and installation of an artificial nest for white storks;
- Study and management of farmland bought, representing 60 ha (in collaboration with the COCA);
- Signing of a land-management agreement with the air force for the conservation of the Marigny aerodrome, previously a NATO base.

The main species concerned are: little bustard, stone curlew, short-eared owl and wheatear. The site is also of significant botanical and entomological interest (E. Mutschler, pers. comm).

The Beauce region
A study has been made on the status and distribution of the little bustard in this region of intensive agriculture, documenting the population's dramatic decline and to see what measures are needed to stabilise its numbers and eventually recolonise deserted sites. The study was proposed by SEPNE and carried out by GEPANA. It should be put into action in spring 1991. The project is being subsidised by the Essonne General Council and the Hunting Federation and financial aid will also be requested from the EC. The results from the study can then be applied to other intensive agricultural areas once they have been adapted to the local context. The measures undertaken should prove beneficial for stone curlew, harriers, quail and grey partridge.

The Lorraine
This region is situated on the migratory path used by cranes each year. Some of them winter between November and April. The aim of the project, which began in 1990 and should end in 1993, is to buy 120 ha of low-lying grassland, part of which has a tendency to flood, to conserve the cranes' resting areas and to ensure the "animal" part of their diet. If necessary these grasslands will be flooded. This project, which has been undertaken by the "Conservatoire des Sites Lorrains" will also prove to be both educational and informative as it will be open to the general public (A. Salvi, pers. comm.).

Prospects
French agricultural plains are today devoted to intensive (almost industrialised) production of cereal crops (since 1950 the annual yield has been quadrupled). The Common Agricultural Policy (CAP) appears unlikely to bring about any significant changes in this practice.

The so-called less favoured areas (LFA) that would be entitled to financial support, tend to be situated in hill areas or areas still covered in grassland (INRA 1989).

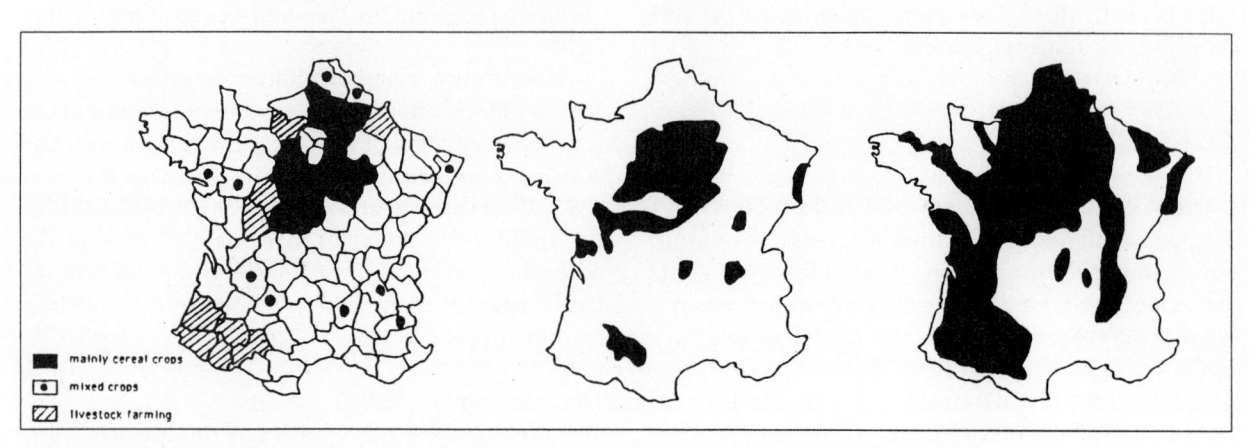

Figure 4: The overall pattern of intensive agriculture in France in 1970, 1979 and 1985

Plate 1: Marigny aerodrome [photo: Arnaud Callec]

The removal from production of agricultural land recommended by the CAP in neighbouring countries already represents a large area. More than 200,000 ha have already been taken out of production in Germany. France only ranks fifth as far as the set aside policy is concerned. Pinault and Lefranc (1990) have asked whether France is prepared to go over the 500,000 ha mark, or maybe even reach the 100,000 ha mark to reassure the EC of her willingness to reduce production.

Low incentives are the main reason for the lack of the enthusiasm shown by farmers and the few who do participate tend to choose to leave their land permanently fallow rather than using crop rotation or grazing fallow: the only farmers concerned are those who are already livestock owners (Pinault and Lefranc 1990). This procedure does not benefit grassland birds as a certain level of land maintenance is necessary to ensure the upkeep of appropriate plant-life.

The introduction of dairy quotas several years ago has meant that many farmers have chosen to rear livestock intensively, indoors, rather than in meadows.

The set aside land can be used for intensive farming or, sadly, often subsequently afforested.

Studies on integrated agriculture (use of biological or physical practices to kill pests rather than chemical products, and the preferential use of natural fertilisers) are being undertaken in many countries of Europe. Some of them already apply new agricultural practices in fields. In France, research is being carried out, but the results will not be available for a couple of years. In 1991, the EC research programme CAMAR (Competitiveness of Agriculture and Management of Agricultural Resources) will begin: in the whole Community, many projects concerning integrated agriculture will be subsidised.

The improvement and extension of biological agriculture is subsidised in the United Kingdom and Germany. In France an institute for techniques in biological application to agriculture has never existed, due to lack of money. Only regional or departmental research units exist.

These two kinds of agriculture are of particular benefit to dry grassland birds thanks to the preservation of the soils' structure and decreasing use of chemical products. However, they will probably not be developed in France in the next few years.

After looking into the status and species distribution of dry grassland birds in France, we can only feel anxious for their future. Their numbers have suffered from a dramatic decline since 1980 and French and European agricultural policy does nothing to curb this trend. The numbers of dry grassland birds risk becoming increasingly small over the next few years and their distribution will certainly become less widespread — contracting into protected ares which remain unaffected by agriculture, and to aerodromes and military bases.

Even if conservation measures become more

important and more widespread they will never be able to compensate for the overall lack of policies to protect dry grassland birds.

CONCLUSION

France has a large variety of landscapes, climates and soils which has until now enabled many dry grassland bird species to inhabit our country for breeding, migration and wintering.

Some of the species in France have reached the northern limit of their distribution. Their populations are low and their numbers will decrease as their habitat disappears. Examples are pin-tailed sandgrouse, lesser kestrel, the pratincoles, calandra and thekla larks. Species such as the little bustard and the stone curlew have found arable farming land to be an alternative habitat. At first, these species adapted to such habitats without problems. However, modern farming methods (such as the increase in farm machinery, modern cropping and use of agrochemicals) had led to an accelerating decline in their populations since the 1960s and 1970s. This decline has been even greater since the 1980s, and even species that had few special ecological requirements are now under threat because their habitats have been altered so drastically.

Most of the species we have looked at in this study are declining rapidly. In some cases, their numbers are becoming so low in our country that the species will be difficult to maintain (e.g. the pratincoles, lesser kestrel). The little bustard may also be in a critical position very soon. The high speed with which farming practices are changing and the more widespread they become, the more likely it is that these species will be under serious threat at the end of the decade, unless agricultural policies are radically altered.

Already the populations of species listed by Goriup and Batten (1990) under "other dry grassland birds in Europe" are declining at an alarming rate. The number of breeding lapwings has diminished by 50 per cent since 1960 and the little owl, which is found in even fewer areas, is also under serious threat. This study has enabled us to obtain an overall picture of the situation of dry grassland birds. It would be interesting to continue this work on a real and statistical basis. In fact, it will be necessary to continue to look at agricultural development to be able to assess its impact on the dry grassland birds and their habitat. The results of studies undertaken concerning set aside and its effect on the land should also reveal some interesting data for preservation of all grassland birds in dry or wet grasslands.

ACKNOWLEDGEMENTS

We should like to thank the following people for sending data about dry grasslands and their avifauna: G. Cheylan, P. Daget, D. Daske, J.F. Dejonghe, C. Girard, J. Garoche, A. Guyot, D. Muselet, E. Mutschler, P. Nicolau-Guillaumet, A. Salvi and J.G. Walmsley.

REFERENCES

Beaufort, de. (1983). *Livre rouge des espèces menacées en France*. Tome 1: les vertébrés. Paris.

Biber, J-P. (1990). Action plan for the conservation of western lesser kestrel *Falco naumanni* populations. *ICBP Study Report* No. 41. International Council for Bird Preservation, Cambridge.

Broyer, J. et al. (1988). *Dépérissement des populations d'oiseaux nicheurs dans les sites cultivés et prairiaux: les responsabilités de la modernité agricole*. SRETIE, Secrétariat à l'Environement, Paris.

Cheylan, G. (1990). Le statut du Ganga cata (*Pterocles alchata*) en France. *Alauda* 58(1): 9-16.

De Ploey, J. (1990). La conservation des sols. *La recherche* No.227.

FIR/UNAO (Ed.) (1982). *L'estimation des effectifs de rapaces nicheurs non-rupestres de France*. Paris.

Goriup, P.D. and Batten, L. (1990). The conservation of steppic birds – a European perspective. *Oryx* 24: 215-223.

Goriup, P.D. and Schulz, H. (1990). Conservation management of the white stork: an international opportunity. *ICBP Study Report* No. 37. International Council for Bird Preservation, Cambridge.

INRA (1989). *Grand Atlas de la France rurale*. Ed. J.P. De Monza, Paris.

Moreau, G. (1990). Une nouvelle espèce nidificatrice pour la France: la Grue centrée (*Grus grus*). *Alauda* 58(4): 244

Muselet, D. (1981). L'Alouette calandrelle (*Calandrella brachydactyla*) et le Bruant ortolan (*Emberiza hortulana*) nicheurs à Pithiviers-le-vieil (45). *Nat. Orléanais* III No. 3.

Muselet, D. (1984). Résultat de 3 années de recensement de l'Alouette calanadrelle (*Calandrella brachydactyla*) dans la Beauce du Loiret. 1982-1983-1984. *Naturalistes orléanais* Vol. 3 No. spécial.

Pinault, M. and Lefranc, J.N. (1990). Gel des terres, jachères de toutes les couleurs. *Agri-décideur*. No. 19: 81-84.

Thonnerieux, Y. (1987). Vanneaux, rien ne va plus. *Revue Nationale de la Chasse* avril 1987: 34-37.

Walmsley, J.G. (1970a). Une Glaréole de Nordmann (*Glareola nordmanni*) en Camargue. Première observation et premier cas de nidification pour la France. *Alauda* 38: 295-305.

Walmsley, J.G. (1970b). Une Glaréole à ailes noirs (*Glareola nordmanni*) en Camargue. *Alauda* 44: 334-335.

Walmsley, J.G. (1988). Nouvelle observation d'une Glaréole à ailes noirs (Glareola nordmanni) en Camargue. *Alauda* 56(4): 430-432.

Walmsley, J.G. (1990). *The breeding status of the two pratincole species (Glareola pratincola and Glareola nordmanni) in France*. Synthèse non publiée réalisée pour ce document.

Yeatman, L. (1976). *Atlas des oiseaux nicheurs de France*. Société Ornithologique de France et Minsitère de la Qualité de la Vie.

STATUS OF LOWLAND DRY GRASSLANDS AND BIRDS IN ITALY

Francesco Petretti

Via degli Scipioni 268/a, 00192 Roma, Italy

ABSTRACT

The situation of dry lowland grasslands in Italy is reviewed and some information on the populations of typical steppe birds is given. The main lowland dry grasslands are distributed in four Italian Regions: Latium, Apulia, Sardinia and Sicily. Although true Mediterranean dry steppes have been drastically reduced due to agricultural changes and urban development, they still cover 200,000 ha which are of outstanding importance for maintaining populations of little bustard *Tetrax tetrax* (2,000-2,500 individuals), stone curlew *Burhinus oedicnemus*, bee-eater *Merops apiaster*, calandra lark *Melanocorypha calandra*, black-eared wheatear *Oenanthe hispanica*, raptors and a significant number of plants and invertebrates.

INTRODUCTION

This paper aims to update the information given in a previous report (Petretti 1988) and to present the results of the first conservation measures undertaken to preserve a sample of the lowland dry grasslands so far identified. Official data were provided by the latest agricultural census in 1982 by Istituto di Statistica (ISTAT) and Istituto Nazionale di Economia Agricola (INEA). These were checked through field surveys and with the help of detailed studies quoted in the references. Field work was carried out from 1982 to 1990 in four regions (Latium, Apulia, Sicily and Sardinia) in the framework of studies sponsored by Unione Nazionale Associazioni Venatorie Italiane (UNAVI) with funds from EC, and by WWF Italy. The major gaps have been covered with additional information provided by ornithologists and conservationists. The bird species concerned are those listed by Goriup and Batten (1990) with the addition of a small group of typical grassland birds of the Mediterranean.

RESULTS

The status and distribution of lowland dry grasslands

Since grasslands in the Mediterranean represent the result of the impact of man and livestock on more complex native vegetative types, it is not easy to identify the main habitats which can be grouped under the definition of "lowland dry grassland". According to official statistics summarized by Angle (1990), in Italy the herbaceous ecosystems used to produce fodder consist of 3,800,000 ha of permanent pastures and 1,150,000 ha of permanent grasslands (Figure 1). The distinction is not clear and most data seem unreliable, due to the broad criteria used to identify the categories. In a previous inventory (Petretti 1988) I also included man-managed grasslands: semi-natural pastures, fallow lands and secondary steppe habitats. These areas have been ploughed at least once in the last three years and have been sown with barley, oats or other herbaceous plants to improve the quality of pastures. Extensive cereal crops and alfalfa fields should also be added to the group of semi-natural and man-managed herbaceous ecosystems, because they are important for bird-life.

For the purpose of the present study, only true grasslands never ploughed are considered. This category comprises 950,000 ha of mountain pastures (from 1,000 to 3,500 metres of altitude in the Appennines and in the Alps), and 205,300 ha of true lowland "steppes". The steppic habitats so far identified include the highest quality habitats with undisturbed ground and rich flora grazed by free-ranging livestock (mainly sheep).

All over the country, lowland steppic habitats, which rise up to 800 m in the Sardinian plateau, share some common features. They are characterized by a Mediterranean climate, with autumn and winter rainfall not exceeding 800 mm (rainfall ranges from 300 mm in the driest area of Salento, Apulia to 800 mm in the wettest area of Tolfa, Latium) and by a rather poor soil with calcareous stones. The maximum average daily temperature, recorded in July, is 25-28°C. The dominant vegetative type is represented by annual grasses (*Poa, Stipa, Alopecurus, Triticum, Hordeum, Avena*), Compositae and perennial plants with succulent roots (asphodels, lilies, irises, orchids, *Ferula, Urginea*). A few areas show scattered bushes (*Rubus, Pyrus, Prunus, Crataegus* and evergreen Mediterranean scrub). A tiny percentage is represented by special vegetative types, which grow on alkaline soils (e.g. *Limonium* and *Suaeda* vegetation of dry mud close to wetlands). Although covering only small surfaces, these habitats are critical for the conservation of invertebrates, plants and bird-life.

The distribution of the main steppe areas is shown in Figures 2 to 5. Apulia accounts for 65,500 ha of steppes (31.9 per cent) and Sardinia for 102,100 ha (49.7 per cent). Smaller areas occur in Latium and Sicily where major agricultural changes took place soon after the Second World War. Sicilian steppe ecosystems (Figure 4) have been identified by mapping the results of the Breeding Bird Atlas (Massa 1985) on the 1:25,000 grid. I selected five true steppe species (stone curlew, roller *Coracias garrulus*, calandra lark, tawny pipit *Anthus campestris*, lesser grey shrike *Lanius minor*) checking the result with the help of satellite imagery and soil maps. The total surface was thus estimated. Although it was not possible to check all the information with field surveys, it is still possible to identify the major ecosystems which are listed in Table 1. The list is an attempt to identify those areas where the creation of nature reserves would be effective for the conservation of steppe vegetation and wildlife.

Status and distribution of steppe birds

Bird-life represents the most striking aspect of the wildlife inhabiting the steppe areas, which provide shelter and food to a rich community of invertebrates, reptiles, amphibians and plants. Table 2 lists the steppe bird species breeding in Italy, adding a few Mediterranean species to the list of Goriup and Batten (1990). From a general point of view, Italian steppes represent a critical habitat for the little bustard, since the Sardinian population, ranging from 1,435 to 2,075 individuals according to Schenk and Aresu (1985), is one of the biggest in Europe (Schulz 1985). Other ecosystems are important breeding areas for the stone curlew, the calandra lark (the population probably exceeds 100,000 pairs), the spectacled warbler *Sylvia conspicillata*, the bee-eater (the population exceeds 10,000 pairs), and the lanner falcon *Falco biarmicus* (in Apulia and in Sicily there are 50 pairs, which represents the biggest number in Europe).

A case study: habitat selection by little bustard males

The little bustard, which can be considered to be the flagship species of steppe habitats, was studied from 1982 to 1990 in Sardinia and in Apulia in order to obtain information on display behaviour, breeding density and habitat selection. I monitored 24 leks of males in Sardinia and 15 in Apulia. Male density ranged from 2.0 males/100 ha to 2.8 males/100 ha in Sardinia (mean = 2.4 males/100 ha ±0.32) and from 1.4 males/100 ha to 2.0 males/100 ha in Apulia (mean = 1.5 males/100 ha ±0.43). The combined mean is 1.9 males/100 ha (±0.58). The habitat composition of the monitored lek area was: 81.3 per

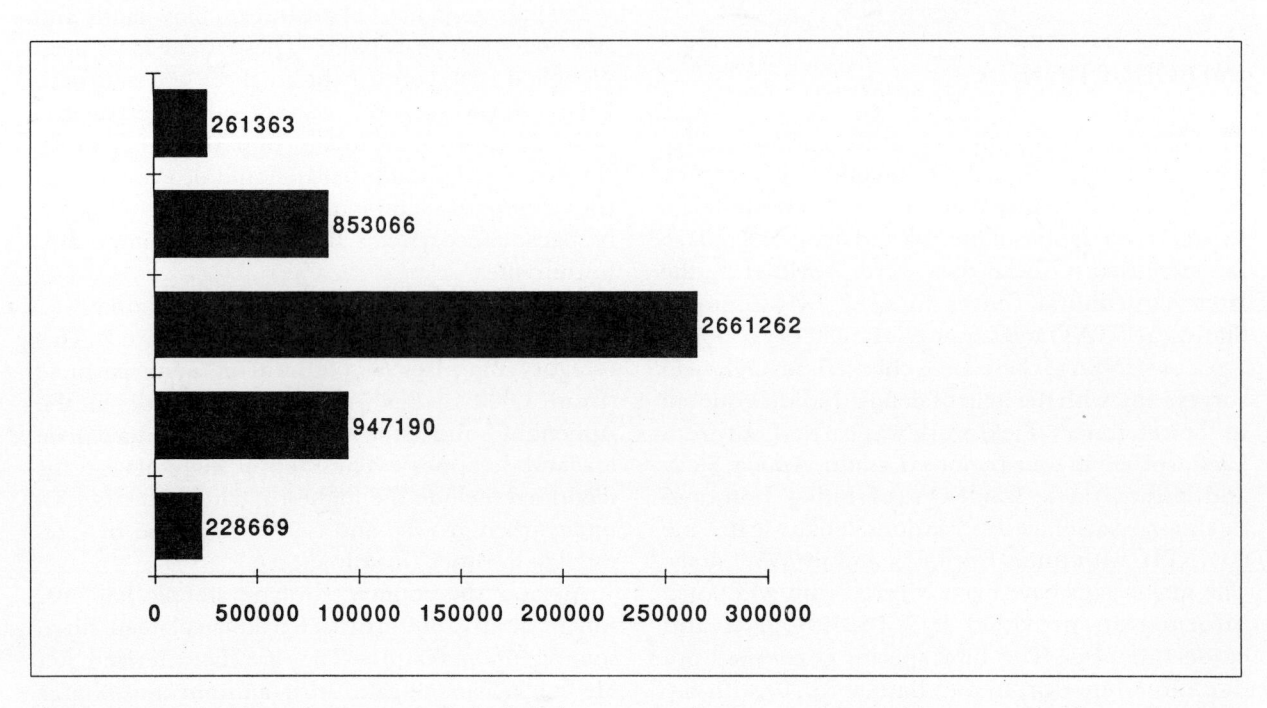

Figure 1: Composition of pastures in Italy (ha): from top to bottom: semi-natural wet grasslands; semi-natural natural dry grasslands; other (fodder cultivations); mountain grasslands; steppes [after Angle 1990]

Plate 1: Little bustard [Panda Photo]

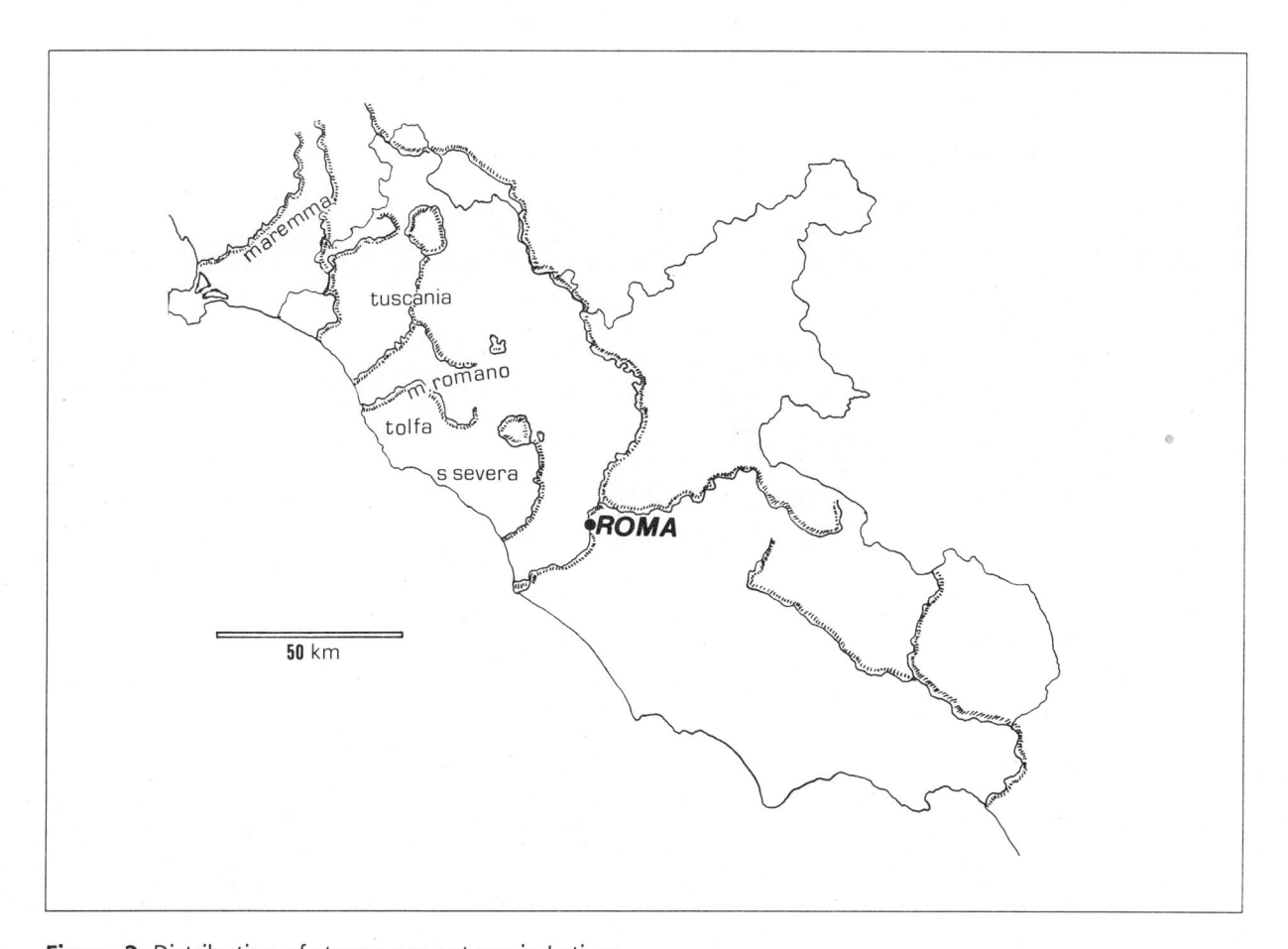

Figure 2: Distribution of steppe ecosystems in Latium

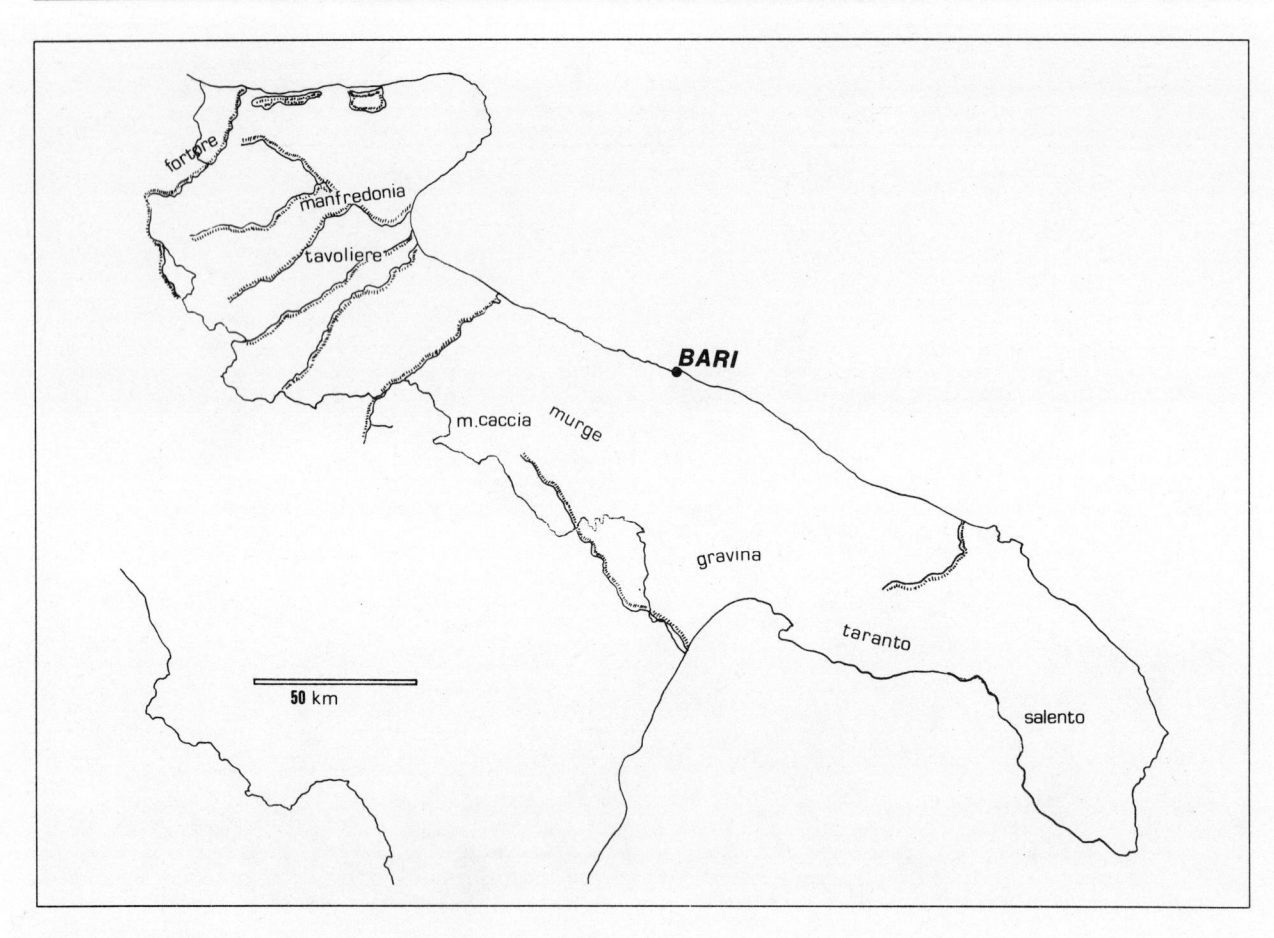

Figure 3: Distribution of steppe ecosystems in Apulia

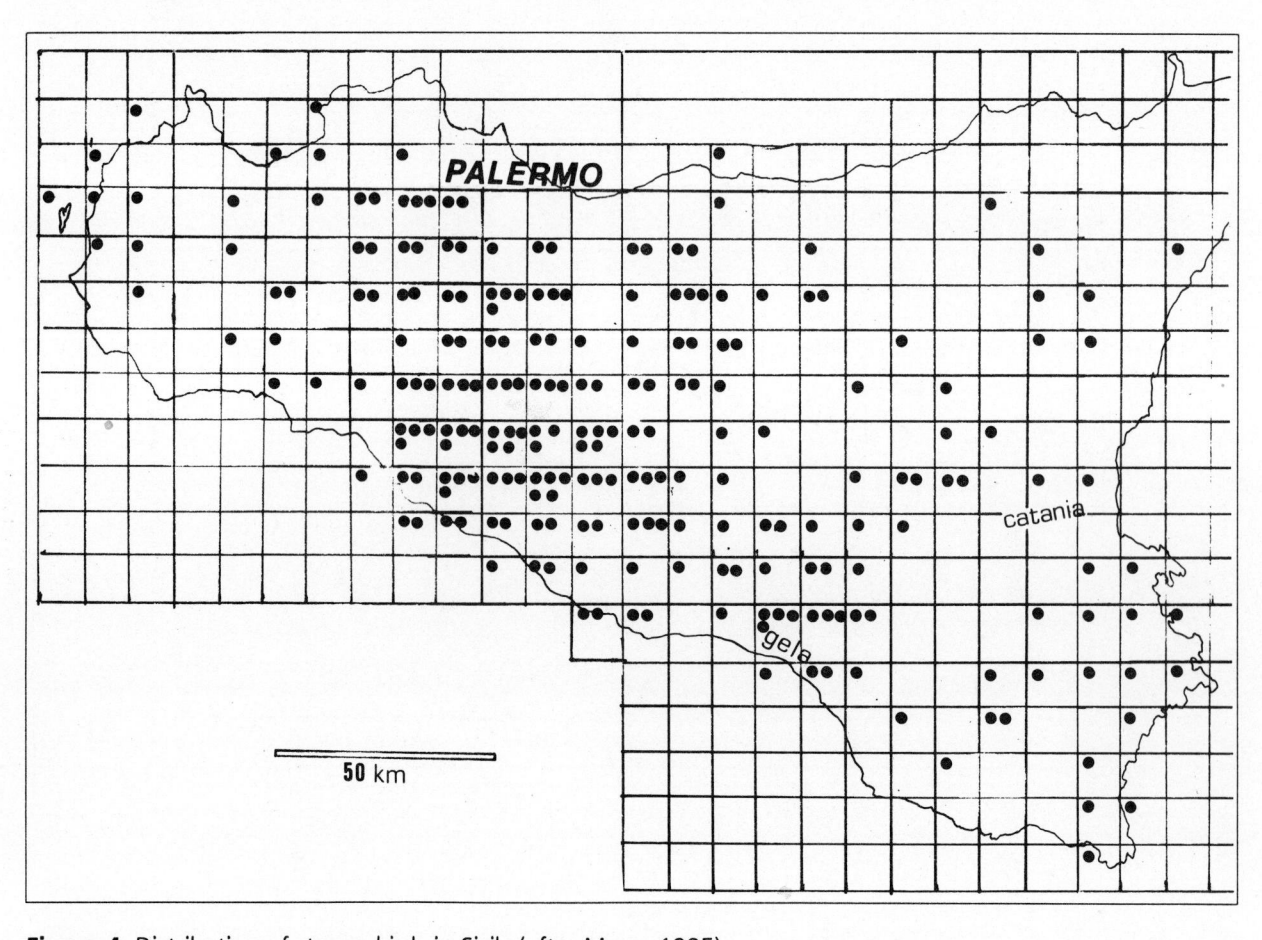

Figure 4: Distribution of steppe birds in Sicily (after Massa 1985)

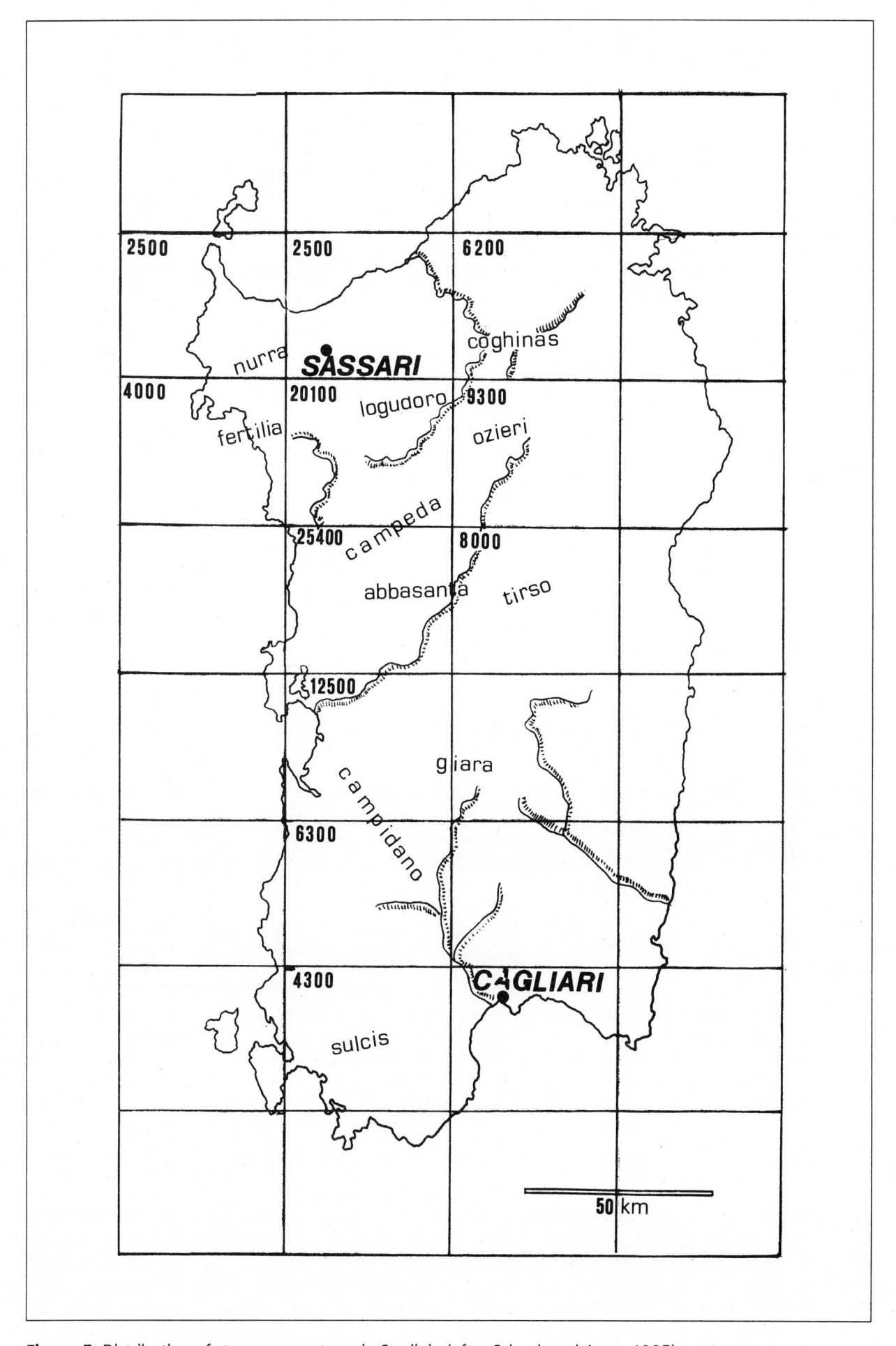

Figure 5: Distribution of steppe ecosystems in Sardinia (after Schenk and Aresu 1985)

Table 1: List of steppe areas in Italy

Region	Denomination	Area (ha)	Notes
LATIUM 27,600	Tolfa 42°N 12°E	25,000	fragmented cattle pastures
	S.Severa 42°N 12°E	100	alkaline steppe
	Monte Romano 42°N 12°E	1,000	cattle pastures in military field
	Tuscania 42°N 12°E	1,000	fragmented sheep pastures in cereal crops
	Maremma 42°N 12°E	500	fragmented sheep pastures in cereal crops
APULIA 65,500	Fortore 42°N 15°E	500	fragmented sheep pastures in cereal crops
	Manfredonia 42°N 16°E (IBA 095)	13,000	fragmented dry stony pastures, cereal crops and garigues
	Tavoliere 42°N 16°E	1,000	fragmented natural and alkaline steppe
	Murge di Bari with M.Caccia 41°N 16°E (IBA 096)	43,200	extensive stony pastures
	Gravina or Murge di Taranto 41°N 17°E (IBA 097)	1,800	fragmented dry stony pastures
	Salento di Lecce 40°N 18°E	6,000	fragmented dry stony pastures
SICILY 10,100 ha	Marsala (IBA 135) 37°N 12°E	100	fragmented alkaline steppes close to coastal marshes
	Capo Feto (IBA 136) 37°N 12°E		
	Simeto (IBA 137) 37°N 15°E		
	Capo Passero (IBA 139) 36°N 15°E		
	Caltagirone, Caltanis setta, Agrigento, Val di Marsala 37°N 13°E	10,000	fragmented sheep pastures in cereal crops
SARDINIA 102,100 ha	Ozieri Logudoro (IBA 112) 41°N 09°E	29,400	extensive stony pastures grazed by sheep
	Altopiano di Campeda (IBA 114) 40°N 09°E	19,500	"
	Upper Campidano Nuradeo (IBA 116) 40°N 09°E	1,000	"
	Tirso e Abbasanta (IBA 11) 40°N 09°E	13,900	"
	Lower Campidano, Giara and Sulcis 40°N 09°E	23,100	"
	Nurra Sassari 41°N 09°E	5,000	"
	Fertilia 41°N 09°E	4,000	"
	Coghinas e Pausania 41°N 09°E	6,200	"
TOTAL AREA		**205,300**	

Key
(*) geographical coordinates approximated to the nearest parallel/meridian
(**) IBA (= Important Bird Area) according to Grimmet and Jones (1989)

cent dry stony pastures, 13.4 per cent oat/barley fields and 5.3 per cent wheat fields. Although little bustard males made some use of all the available herbaceous ecosystems (including intensive cereal crops) it is clear that they showed a strong preference for permanent pastures grazed by cattle and sheep.

DISCUSSION

Over the course of centuries, steppe wildlife has shown a large degree of adaptation to human activity, consisting mainly of seasonal grazing by sheep and other free-ranging livestock. The seasonal cycle of grazing is well documented in Latium and Apulia where the natural pastures are visited by sheep and shepherds mainly in the cold season (autumn and winter), through a migration called *transumanza* from and to summer mountain grasslands in the Appennines. This movement, which occurred in the past by foot and now by truck, is still a typical feature of the sheep economy in continental Italy, although residential stock farming is spreading thanks to the improvement of fodder production in irrigated pastures. Thus the conservation of steppe habitats depends not just on the sheep economy, but actually on extensive sheep farming.

Permanent pastures on dry soils and herbaceous ecosystems not actively managed by man still represent a viable proportion of farmland in Italy but their area is being reduced by agricultural improvement and expanding towns (Figure 6). On the other hand, the areas still suitable for birds have been progressively fragmented and the common pattern is now a mosaic of small "islands" of steppe scattered among intensive cereal crops. This proved to have deleterious effects

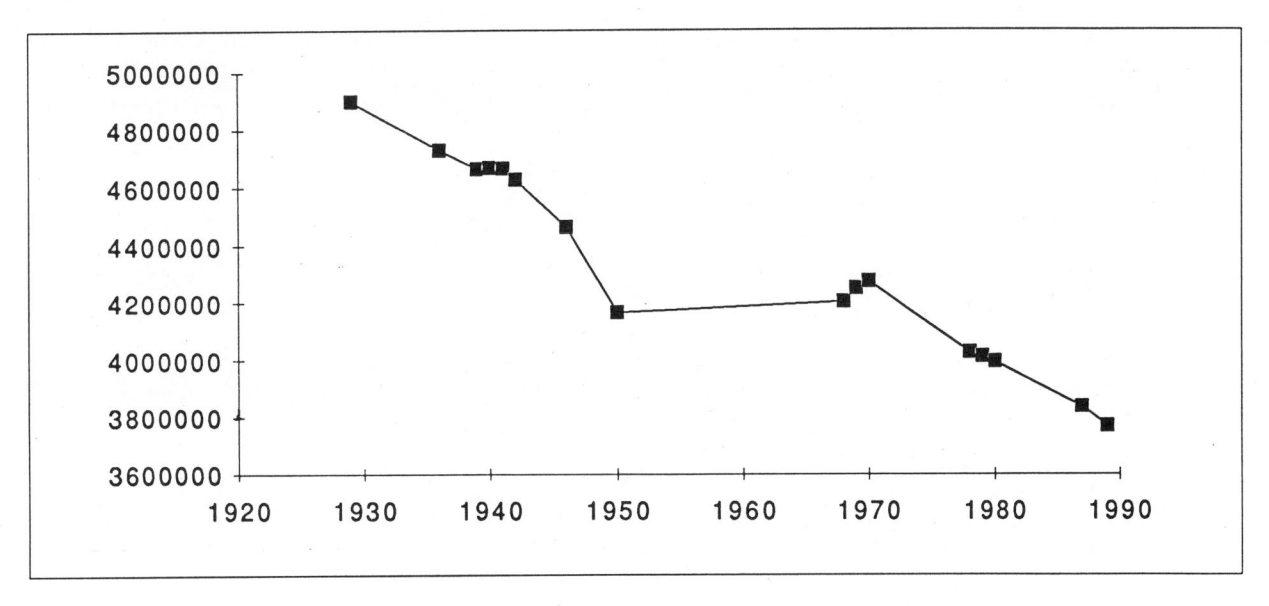

Figure 6: Decrease in total area of pastures in Italy (after Angle 1990)

on the largest and most mobile and shy species of birds as shown by the case of the little bustard in Apulia. Male birds were ousted from very good habitats due to the construction of roads and houses with the attendant human disturbance.

The application of set aside rules does not seem to be enforced in Italy at the moment. Most farmers apply to the Ministry of Agriculture to get set aside compensations just for the usual fallow period of the cereal crop cycle, which is something very different from permanent or semi-permanent uncultivated areas. Many grants are still available to transform seasonal grazing *transumanza* into stable and more intensive forms of stock farming.

Large irrigation schemes will provide water for agriculture in dry areas around the Coghinas lake in Sardinia and the Tolfa and Santa Severa grasslands in Latium. The result will be the development of vineyards, kiwi-fruit plantations and vegetable cultivations in greenhouses particularly in the areas served by the Cassa del Mezzogiorno which is a funding mechanism set to distribute grants in Southern Italy. Attention should also be paid to the Integrated Mediterranean Programme (IMP) and Common Agricultural Policy (CAP) which in Italy seem to fight with the environmental regulations set by the EC.

The greatest difficulty faced by conservation organizations seems to be represented by the lack of the status of "natural place" for the grassland ecosystems: they are still treated as second-class farmland. In the recent past it was the same for wetlands which are now considered as places of great economic value and ecological importance. A campaign to implement a convention for the preservation of dry lowland ecosystems could play a major role for the conservation of these habitats in EC countries. A preliminary step was the inclusion of some grassland ecosystems in the IBA report and the inclusion of Campeda plateau in the list of the five Italian places which require priority funds from the EEC.

Table 2: Breeding birds of lowland dry grassland in Italy

Montagu's harrier	*Circus pygargus*
Lesser kestrel	*Falco naumanni*
Quail	*Coturnix coturnix*
Grey partridge	*Perdix perdix*
Little bustard	*Tetrax tetrax*
Stone curlew	*Burhinus oedicnemus*
Collared pratincole	*Glareola pratincola*
Barn owl	*Tyto alba*
Little owl	*Athene noctua*
Hoopoe	*Upupa epops*
Bee eater	*Merops apiaster*
Lesser grey shrike	*Lanius collurio*
Calandra lark	*Melanocorypha calandra*
Short-toed lark	*Calandrella brachydactyla*
Crested lark	*Galerida cristata*
Skylark	*Alauda arvensis*
Tawny pipit	*Anthus campestris*
Spectacled warbler	*Sylvia conspicillata*
Fan-tailed warbler	*Cisticola juncidis*
Stonechat	*Saxicola torquata*
Wheatear	*Oenanthe oenanthe*
Black-eared wheatear	*Oenanthe hispanica*
Linnet	*Carduelis cannabina*
Ortolan bunting	*Emberiza hortulana*
Cirl bunting	*Emberiza cirlus*
Black headed bunting	*Emberiza melanocephala*
Corn bunting	*Miliaria calandra*
Rock sparrow	*Petronia petronia*

REFERENCES

Angle, G. (1990). *Habitat: guida alla gestione degli ambienti naturali*. Ministry of Forestry and Agriculture and WWF Italy, Rome.

Goriup, P. and Batten, L. (1990). The conservation of steppic birds—a European perspective. *Oryx* **24**: 215-223.

Grimmett, R.F.A. and Jones, T.A. (1989). *Important bird areas in Europe*. ICBP Technical Publication No. 9. International Council for Bird Preservation, Cambridge.

Massa, B. (Ed) (1985). *Atlas Faunae Siciliae - Aves.Il* Naturalista Siciliano, vol.IX.

Petretti, F. (1988). An inventory of steppe habitats in southern Italy: 125-143. *In: Ecology and conservation of grassland birds"*, ICBP Technical Publication 7. Cambridge.

Schenk, H. and Aresu, M. (1985). On the distribution, number and conservation of the Little bustard in Sardinia (Italy), 1971-1982. *Bustard Studies* 2: 161-164.

Schulz, H. (1985). A review of the world status and breeding distribution of the Little bustard. *Bustard Studies* 2: 131-152.

STATUS OF LOWLAND DRY GRASSLANDS AND GREAT BUSTARDS IN AUSTRIA

Hans Peter Kollar

Kirchengasse 34, 2285 Leopoldsdorf, Austria

INTRODUCTION

The east of Austria lies in the range of the pannonic climate and is part of the flora province *Pannonicum*. The climax vegetation is in most of its range thermophilic oak woods that in the wide open lowlands would be (and were) developed as sparse wooded steppe (Wendelberger 1954, 1989; Niklfeld 1964). After a postglacial phase of warming, when the open parts of eastern Austria again were covered with wood, these sparse oak wood steppe habitats were certainly influenced by human activity as early as the Bronze Age and presumably also kept open. Primary steppes in our region as well as in Hungary were edaphic, restricted to, for instance, the tops of loess hills, sandy and rocky soils (Soo 1940).

LOWLAND DRY GRASSLAND DISTRIBUTION

The pannonic east of Austria was characterized by the sand steppe habitats of the Marchfeld (east of Vienna), the loess steppe of the *Weinviertel* north of the Marchfeld, and the grass-dominated xerothermic limestone-hill vegetation of the limestone cliffs in the Danube basin and in the Weinviertel. In addition, since earliest times there were secondary dry grasslands, especially pastures, which developed through human impact. As all of the open primary grasslands were restricted to few edaphically determined sites even in the distant past, today they have shrunk down to several small isolated sites (see Holzner 1986, Figure 1).

Most of the relict dry grasslands in the pannonic fauna and flora provinces of Austria are situated in elevated country or in higher altitudes (eastern Alps, limestone mountains near Hainburg, lime cliffs of the *Weinviertel*, Figure 1). The only extensive dry grassland sites remaining in the lowlands are the nature reserves of the sandhills near Oberweiden and the *Stipa*-grasslands known as *Weikendorfer Remise*, both in the Marchfeld, east of Vienna (see Wolkinger *et al.* 1986). The first site is an area of 115 ha consisting of fixed stationary sand dunes, the highest of them 8 m, that carry dry grassland vegetation. A great part of the dunes and their vicinity is forested with conifers. The second site of slightly larger extent is pannonic dry grassland (*Stipa*-grasslands with juniper *Juniperus*) in the Marchfeld, divided between two nature reserves (183 and 37 ha) and surrounded by forestry plantations and farmland. These sites together with some smaller ones are but a shadow of the former wide open grasslands and wood steppe habitats.

THE GREAT BUSTARD

The great bustard *Otis tarda* is one of the most characteristic species of the pannonic lowlands. Up to this century the great bustard lived in virtually all of the predominantly open parts of the landscape in the pannonic east of Austria (Lukschanderl 1971, Winkler 1973). As throughout its range, it had switched from its primary habitat, the grass steppe amply supplied with herbs, to wide, open agricultural land that lay within the ecological range of the species. The current relics of the pannonic primary (or secondary) habitats in Austria mentioned above no longer meet the ecological needs of the great bustard because of their small size and, as they are surrounded by woods, are not open enough. So in Austria the great bustard occurs exclusively on agricultural land.

The development of modern agriculture that was especially severe in the great plains with their fertile soils is, apart from hunting up to 1969, the main reason for the decline of the great bustard population (Winkler 1973, Lütkens 1974, Kollar 1988a for example). Since it is neither possible at present nor realistic in the near future to restore large areas of

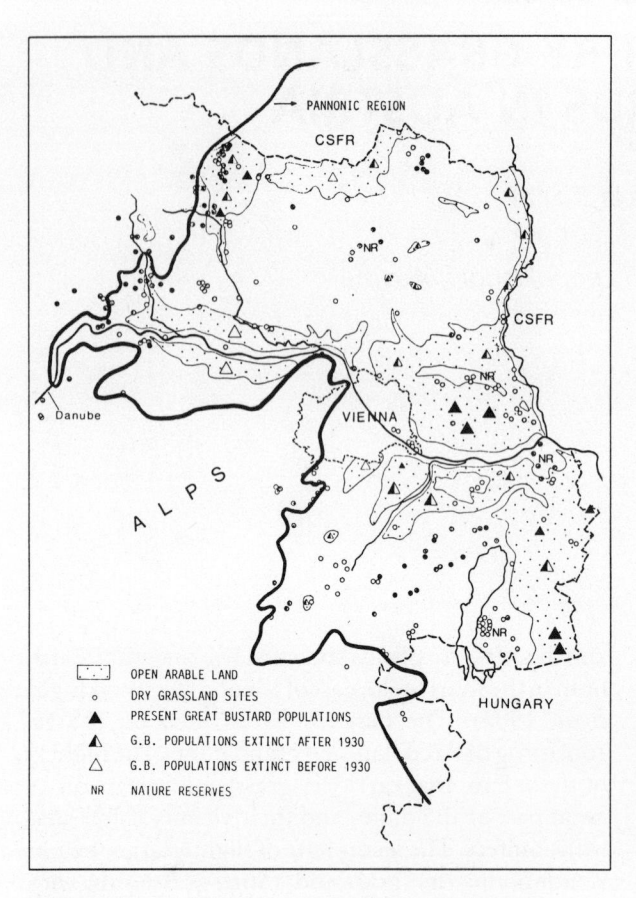

Figure 1: Distribution of lowland dry grassland in Austria

appropriate habitat, (i.e. wide open to lightly wooded grassland), measures for species conservation must remain within the prevailing system of land use and adapt to specific local demands. This is the more urgent as world-wide attempts to strengthen weakened populations with animals raised in captivity have proven unsuccessful (Kollar 1988a). Moreover, studies in areas where semi-natural grasslands and agricultural plots are situated close together in great bustard ranges have revealed that the birds preferred the latter as nesting sites and the former as display and resting places (Farago 1981). In this way they get into the ecological "trap" that is responsible for the reduction of their reproductive success.

In Austria, therefore, habitat management in the broadest sense of the word is carried out. In the main sites used by the great bustard, special plots are set up and managed to meet the demands of the species in respect of breeding sites and resting areas, but which do not try to imitate the original steppe habitat.

In the Marchfeld east of Vienna, so-called bustard plots have operated since 1979. These are areas of about 2-3 ha each, which are leased all year round, tilled with a special composition of seeds and managed accordingly (Lütkens 1974, Kollar 1983, 1988a, 1988b). For the last ten years or so in the *Weinviertel* in the north of Austria, where a cross-frontier Austrian-Czechoslovakian population lives, clover and lucerne crops have been cultivated specifically in the bustard range with support from official authorities. Since 1985, bustard plots corresponding to those in the Marchfeld (Kollar 1988a) have also been established.

Finally, within the range of the Hungarian-Austrian bustard population, initiatives are in hand to cultivate certain plots close to the existing meadow nature reserve as bustard breeding and resting grounds, to be designated as set aside areas with official support (Reiter 1989). The great bustard population in the Marchfeld east of Vienna, which numbered about 120 individuals after the Second World War, has now shrunk to 25-28 individuals (Kollar 1989). Since the early 1980s, the decline has somewhat slowed. Observations on the use of the bustard plots by the great bustard suggest that they have played a part in arresting this decline. The Czechoslovakian-Austrian populations have been decreasing fairly steadily for some decades, and the *Weinviertel* population now comprises about 12-18 individuals. Here, as well, a positive influence of the lucerne and clover cultures and the bustard plots is probable. The habitat-orientated initiatives in the Burgenland at the border with Hungary have up to now been fundamentally impaired by the fact that even in the fully protected core area the meadows have been mown at the time of breeding (Triebl 1988). Only recent intensive activities in connection with the set aside programme made a postponement of the mowing date possible (though in a far from efficient form, Triebl, pers. comm.). A new organisation for the protection of the great bustard in Austria (see below) should help to find long term and permanent solutions here as well.

Table 1: Status of Austrian lowland dry grassland birds

Species	Migrant status	Population size	Population trend	Austrian Red Data Book Status
Little bustard *Tetrax tetrax*	no recent sightings	0	-	Extinct since 1921
Great bustard *Otis tarda*	local breeder	80-90 individuals	-	Endangered
Montagu's harrier *Circus pygargus*	local breeder, migrant	3-7 pairs	-	Endangered
Quail *Coturnix coturnix*	scattered breeder, migrant	?	-	Endangered
Stone curlew *Burhinus oedicnemus*	very rare breeder	<10 pairs	-	locally Extinct, Endangered
Short-toed lark *Calandrella brachydactyla*	rare migrant	0	-	Extinct since 1966
Woodlark *Lullula arborea*	rare breeder	<10 pairs?	-	locally extinct, Endangered
Tawny pipit *Anthus campestris*	scattered breeder	10-20 pairs	-	Endangered

Plate 1: Managed plot for great bustard in the Marchfeld, Austria [photo: Hans Peter Kollar]

In the meantime, the Austrian part of this population has gone down to about 25 individuals. Together with some smaller populations, at present a total of about 80-90 great bustards live in Austria.

Measures for the conservation of the great bustard in Austria since the Second World War have been carried out by individuals and institutions, which have been supported since around 1980 by government authorities, industry and hunting and nature conservation associations. In 1990, an all-Austrian "community of interests" was formed, that will better coordinate the measures, facilitate communication and shape the distribution of more efficient conservation initiatives for the great bustard.

OTHER DRY GRASSLAND BIRD SPECIES

Of the threatened and rare species of ground-nesting birds in continental Europe listed by Goriup (1988), eight are included in the Austrian Red Data Book (Gepp 1983, Spitzenberger 1988). All of them are endangered or extinct (Table 1).

THREATS TO DRY GRASSLANDS

Not only from the point of view of the conservation of the great bustard, the restoration of species-rich green countryside (from narrow corridors to open (pasture-) land), is one of the most urgent demands of today. Apart from this, the conservation and management of existing primary and secondary dry grasslands is of highest priority since these habitats are threatened in many ways (following Holzner 1986): afforestation, agricultural use, building activity, fertilization, intensive use for recreation, burning, consolidation, refuse deposition, planting of *Robinia* (in the pannonic climate) and other trees, quarrying etc.

Conservation measures

These threats are valid not only for Austria but certainly for other European countries as well. Therefore, the main measures for conservation of dry grasslands seem in general to be:

- Pure conservation (certain types of primary dry grasslands that develop without human interference), setting up nature reserves;
- Conserving dry grasslands of all sorts, including narrow border habitats, slopes and small patches;
- Giving grasslands priority status in habitat assessments during nature-relevant legal proceedings, farm management plans etc.;
- Sympathetic management by mowing, grazing and other forms of "extensive" use;
- Elimination of disturbance (for example spatial and temporal channeling of recreation

pressure, shielding and buffering against sources of disturbance, such as roads);
- Enlargement and compounding of corridors and patches of semi-natural landscape elements including dry habitats of all kinds, rethinking of farmland intensification and its methods, aims and purposes.

Current and future trends

In Austria currently, efforts are being made to develop programmes and models for the furtherance, support and management of set aside plots, meadows, grasslands and border habitats (see for example Manzano 1989). Apart from these recent initiatives and administrative structures, special species conservation programmes are being carried out which also include the conservation and management of certain areas, for example, for the white stork *Ciconia ciconia* (Kirchberger 1989b) and the great bustard (see above and Kollar 1983, 1988a,b, Kirchberger 1989a).

Habitat-orientated species conservation programmes are in our opinion a useful means to restore and maintain secondary grasslands. Methods are straightforward and are (largely) without administrative impediments (e.g. the use of simple leases). But of course they cost money which in most cases has to be raised by nature conservation organisations, private societies, hunting associations or other interested bodies. Efforts and administrative structures that are carried and secured by (federal) authorities and based on legal regulations and proceedings, perhaps in coordination with international associations, would put the restoration of secondary grassland as well as primary grassland on a new basis.

The sites of dry grassland in Austria have recently been listed and published in a catalogue (Holzner 1986). Although public awareness of the value of dry habitats is low compared to wetlands (like the riverine forests that are much less "untouched wilderness" than most of the dry grasslands), this and similar publications will help to reduce the danger of uncontrolled ("over-night") destruction of these sites. In addition, some scientific management programmes have been in progress for many years (Waitzbauer 1990). For most of these sites of primary dry grassland, simple conservation (nature reserves), elimination of disturbance and consistent management uder supervision of the relevant scientists are the measures to be taken to preserve them.

REFERENCES

Farago, S. (1981). Összehasonlizo Mikroklima-viszgalatok a tuzok (Otis tarda L.) Hansagi Feheszkelöhelyen. Vadbiologiai Kutatas 27: 25-32.

Gepp, J. (ed.) (1983). Rote Listen Gefährdeter Tiere österreichs. Grüne Reihe des Bundesministeriums für Gesundheit und Umweltschutz. Wien. 242pp.

Goriup, P.D. (1988). The Avifauna and Conservation of Steppic Habitats in Western Europe, North Africa and the Middle East. *In:* Goriup, P.D. (ed.) (1988). Ecology and Conservation of Grassland Birds. ICBP Technical Publ. No.7, pp.145-157.

Holzner, W. (ed.) (1986). Österreichischer Trockenrasen-Katalog. Grüne Reihe des BMfGU band 6, Wien. 380pp.

Kirchberger, K. (1989a). Naturschutzarbeit und Landwirtschaft. Club Niederösterreich 2/1989: 74-79.

Kirchberger, K. (1989b). WWF Weißstorch-Schutzprojekt Rust. Vogelschutz in Österreich 4: 33-37

Kollar, H.P. (1983). Der Einfluß von Trappenschutzfeldern auf den Aktionsraum der Großtrappe (Otis tarda L.) im Marchfeld (Niederösterreich). Egretta 26(2): 33-42.

Kollar, H.P. (1988a). Arten-und Biotopschutz am Beispiel der Großtrappe (Otis tarda L.). Umwelt, Schriftenreihe für ökologie und Ethologie, Wien. 56pp.

Kollar, H.P. (1986b). Artenschutzprogramm Großtrappe. Vogelschutz in österreich 2: 63-67.

Lütkens, R. (1974). Trappenschutz im Marchfeld. Der Anblick 1974(3): 71-75.

Lukschanderl, L. (1971). Zur Verbreitung und ökologie der Großtrappe (Otis tarda L.) in österreich. J. Orn. 112: 70-93

Manzano, C. (1989). Distelverein – Naturschutz und Landwirtschaft. Club Niederösterreich 2/1989: 8-21.

Niklfeld, H. (1964). Zur zerothermen Vegetation im Osten österreichs. Verh. Zool.-Bot. Ges. Wien 103-104: 152-181.

Reiter, A. (1989). Grünbrachen-Förderungsprogramm im "Wasen"- Hoffnung für das überleben der Großtrappe? Vogelschutz in Österreich 3: 23-28.

Soo, R. von Bere (1940). Vergangenheit und Gegenwart der pannonischen Flora und Vegetation. Nova Acta Leopoldina NF 9/56: 3-49.

Spitzenberger, F. (ed.) (1988). Artenschutz in österreich. Grüne Reihe des Bundesministeriums für Umwelt, Jugend und Familie Band 8, Wien. 335pp.

Triebl, R. (1988). Die Großtrappe im und um das Vollnaturschutzgebiet Hansag (Waasen) - Burgenland. Merkblatt, Landesgr. Burgenland des Österr. Naturschutzbundes.

Waitzbauer, W. (1990). Die Naturschutzgebiete der Hundsheimer Berge in Niederösterreich. Entwicklung, Gefährdung, Schutz. Abh. Zool.-Bot. Ges. Österreich 24: 1-88.

Wendelberger, G. (1954). Steppen, Trockenrasen und Wälder des pannonischen Raumes. Angew. Pflanzensoz. (Festschrift E. Aichinger) 1: 573-634.

Wendelberger, G. (1989). Zur Klärung des Waldsteppen – Begriffes. Ein Versuch. Verh. Zool.-Bot. Ges. österreich 126: 185-195.

Winkler, H. (1973). (Otis tarda Linne 1758) — Die Großtrappe. Beiträge in: Glutz von Blotzheim, U., Bauer, K. und E. Bezzel, Handbuch der Vögel Mitteleuropas, Bd. 5, Frankfurt: 649-688.

Wolkinger, F., Gepp, J., Plank, S. und Zimmermann, A. (1986). Die Natur-und Landschaftsschutzgebiete österreichs. Wien. ÖGNU-Eigenverlag, 255pp.

CONSERVATION OF LOWLAND DRY GRASSLAND BIRDS IN HUNGARY

Ferenc Márkus

Department for Nature Conservation, Ministry for Environment and Regional Development, 1011 Budapest, Hungary

ABSTRACT

Grasslands, such as steppes, represent unique ecosystems in Hungary. These areas are utilized by man for grazing livestock in a traditional manner. Today, however, this farming method is becoming a relic of the past and grassland areas have been shrinking at an alarming rate. As a result, the flora and fauna found only on these habitats have suffered a dramatic loss. At the same time, traditional shepherding skills are doomed.

The Hungarian *pusztas*, parts of which are preserved in their original form and parts formed as a consequence of grazing in previous centuries, are considered to be of a most precious natural heritage in a European context. The threats these dry grasslands and their birds are under, continue due to ever-increasing human disturbance.

INTRODUCTION

Lowland habitats, mainly grasslands, have been drastically altered over the last 40 years by intensive agriculture and a growing economic infrastructure. These habitats were previously affected to a similar extent only in the 18th and 19th centuries when the regulation of rivers, drainage of marshes and deforestation radically reshaped the landscape. That era opened up the possibilities of extensive agriculture.

Changes in agricultural methods from 1950 onwards have resulted in the development of intensive farming along with an expansion in industry and the economy as a whole. The sole aim of these agricultural developments was to increase production. The dramatic decrease in habitat and bird-life was not taken into consideration. Hungary's legendary endless grassland, the *puszta*, is a secondary man-made habitat (Plate 1). It was shaped over the centuries by deforestation, regulation of rivers, drainage of marshes and year-round grazing of livestock. This regular grazing of cattle, horses and sheep ensured the long-term maintenance of this secondary habitat. Of all the habitats of the Great Plain, grasslands and their birds have been the most seriously affected by intensive agriculture over the last 40 years.

The protection of lowland grassland birds began in the early 1970s with site (covering 121,413 ha) and species (33 of those listed by Goriup and Batten, 1990) protection. Unfortunately, there are no current measures for the conservation of unprotected grasslands.

CHANGES IN LOWLAND DRY GRASSLANDS

The changes in natural lowland habitats are somewhat obvious, but they can be measured only indirectly. In the last 40 years the monitoring of environmental conditions, especially that of certain habitats, was almost completely ignored. As a matter of fact, only fluctuations in agricultural statistics, types of cultivation and, in particular, the changes in woodlands were recorded properly. In most cases, on the basis of these data, changes in natural and semi-natural habitats and the causes of these changes can be identified only in an indirect way (Grimmett and Jones 1989). The habitats where the most significant losses have occurred in the last 40 years are (i) natural wetlands; (ii) grasslands; and (iii) hedgerows and forest belts. These losses, from an environmental point of view, have not yet been recorded and analysed. No complete record, neither on a nationwide nor local level, has been made of the habitats and the rate of (and reasons for) their loss.

In 1950 there were 1,474,000 ha of grassland in Hungary, in 1990 1,185,570 ha. That is, in 40 years nearly 300,000 ha (20 per cent) have been lost. The distribution of steppe-like habitats is shown in Figure 1, with the sites listed in Table 1. However, in this

Plate 1: Landscape of the Hungarian *puszta*

Plate 2: Traditional grassland management with grey cattle in the Hortobágy National Park

Plate 3: Hereford cattle and fences on the Hungarian lowland, detrimental to important bird habitats

[photos: F Márkus]

period the degree of qualitative loss was even higher than that of the quantitative loss. For example, while in 1950 within the grasslands the proportion of meadows to pastures was 40:60, by 1990 this proportion has changed to 20:80. Moreover, the 600,000 ha of statistically registered meadows have been reduced to 200,000 ha; a loss of 70 per cent (Figure 2).

A statistical analysis of the the data from the last 40 years leads to the conclusion that losses of grassland occurred only in the case of meadows, while pastures increased slightly. In reality, however, major losses occurred in both cases, although there is no doubt that meadows and their wildlife suffered the most serious decline. Which factors of intensive agriculture caused the losses of meadows? The most serious and basic reason is water drainage. Wide-scale state-subsidised schemes began in the 1960s and resulted in large areas being drained. This led to the degradation of meadows, followed by a decrease in grass-yield. Areas which habitually dried out were ploughed up or used as pasture, which on the one hand increased the statistical proportion of pastures, but on the other caused the further degradation of the former hayfields. Thus, "grasslands" existing in this condition failed to function economically or ecologically (Enyedi 1964). The drained meadows lost their hay yields and their soil and vegetation deteriorated still further as a result of intensive trampling due to the introduction of grazing animals. The original ecological composition of these areas has been irreversibly changed.

During the last 40 years, the majority of the large continuous pastures in Hungary have been lost. The main reasons are:

- ploughing up of areas due to changes in intensive agricultural systems
- creation of fish ponds and paddy fields
- planting of vineyards and orchards
- afforestation (poplar, pine and acacia)
- development of industrial infrastructure (roads, factories and settlements)

Table 1: Protected and unprotected areas (>500 ha) of steppic habitats in Hungary

Name of area	Area (ha)	Region	Conservation Status
Hortobágy (1973)	52,000	SK, HB	NP, BR, RS
Kiskunsági NP (1975) (I-VI)			
Kiskunsági Puszta (II)	11,030	PT, BK	NP, BR
Bugac (VI)	10,920	BK	NP, BR
Szikes Tavak (III)	3,903	BK	NP, BR, RS
Fülöpházi Homokbuckák (IV)	1,665	BK	NP, BR
Pusztaszer (1976)	22,151	CK	LPA, RS
Borsodi Mezoség (1989)	9,168	BAZ	LPA
Hajdusági Erdospuszták (1989)	6,243	RB	LPA
Dévaványa (1975-1990)	5,900	BS	LPA
Agota Puszta (1973)	4,700	HB	NCA
Pitvaros (1989)	3,156	CD, BS	LPA
Orgovány (1976)	2,953	BK	LPA
Harta-Akasztó	2,500	BK	-
Ecseg Puszta (1984)	2,000	SK	LJ
Biharugra (1989)	2,000	BS	LPA
Ferto-tó (1977)	1,700	GN	LPA, BR, RS
Tiszaluc-Kesznyéten	1,700	SK, HB	—
Füzesgyarmati Puszta	1,200	BS	—
Mezogyámi Puszta	1,200	BS	—
Cserebökény Puszta	1,000	SK, BS	—
Csabacsüd	1,000	BS	—
Kunpeszéri Legelo	1,000	PT, BK	—
Füzesgyarmat-Balkán	900	BS	—
Zsákai Puszta	800	HB	—
Karcag-Kunhegyes Puszta	700	SK, HB	—
Soltszentimrei Puszta	600	BK	—
Bucsa-Jeno-major Puszta	600	BS	—
Balmazú jvárosi Nagy-Szik	600	HB	—
Békéssámsoni Puszta	600	BS	—
Polgár Nagykopasz Puszta	500	HS	—

NP = National Park
LPA = Landscape Protection Area
NCA = Nature Conservation Area
LJ = Protection of Local Importance
BR = Biosphere Reserve
RS = Ramsar Site

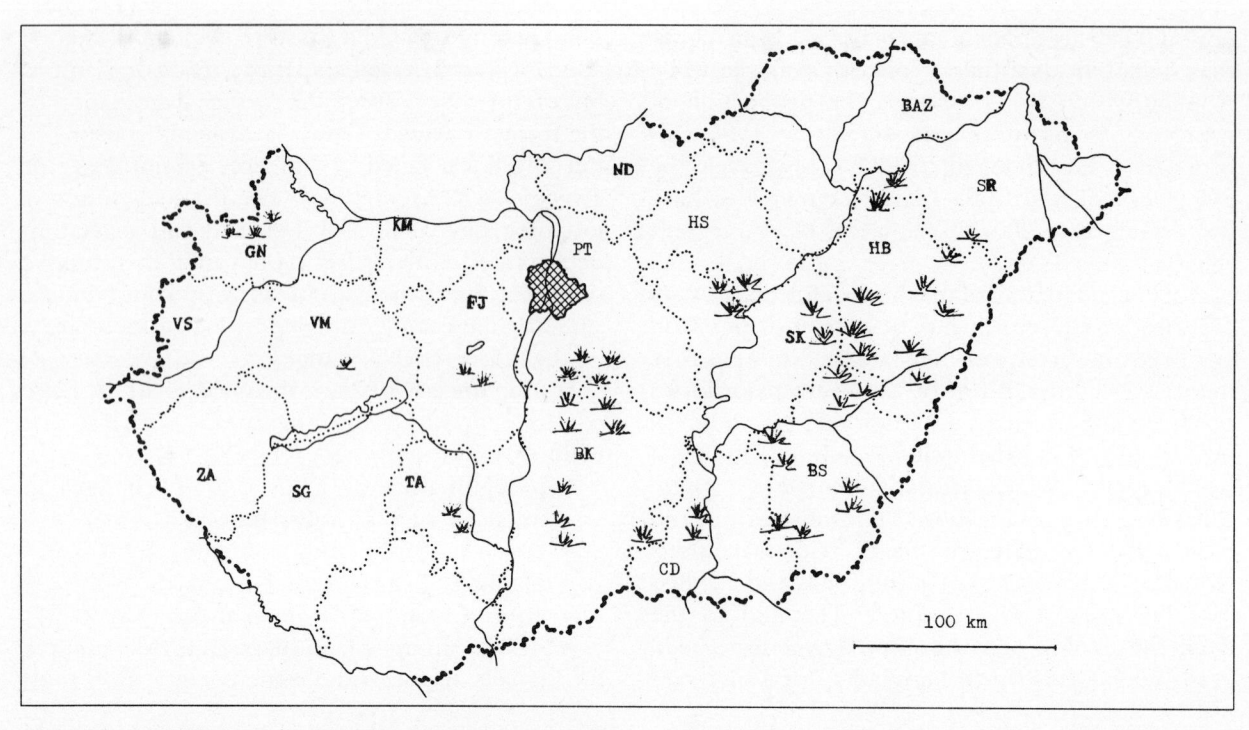

Figure 1: Distribution of steppe-like habitats in Hungary

A contradictory fact seems to be that according to official statistics, pastures have slightly increased in number rather than decreased in the period under discussion. However, it is useful to ask: where does this increase originate? This statistical increase comes from pastures "developed" from drained, degraded meadows and poor, abandoned, uncultivated ploughed fields. The natural value of these recently created pastures is far less than that of pastures created some decades before (Sterbetz 1975). Due to overgrazing the vegetation and soil of these drained meadows are becoming increasingly degraded. They fail to function either as meadows or as dry pasture. Uncultivated areas, such as ploughed fields, have little significance as natural habitat.

The natural bio-diversity of the flora and fauna only comes to life with the aid of correct grazing regimes (Bodrogközy 1970). The number of grazing livestock has decreased significantly over the last 40 years. A lack of permanent grazing, trampling and manuring leads to a general decline of habitat value. To create biodiversity on pastures it is essential to work out optimal grazing methods, numbers of livestock and selection of species and breeds.

Another problem, from an environmental point of view, is that present-day pastures are scattered in small, isolated units which has resulted in the isolation of species. The former agricultural policy insisted on a reduction in the decrease of ploughed fields and an increase in woodland. The loss of ploughed fields and woodland to industry and the like was compensated for at the expense of pasture. The cultivation of most of the remaining isolated pasture ceased within a few years and in this way, due to a lack of grazing animals, it became infested with weeds. As a result of afforestation, sandy-soiled pastures, rich in species, became pine *Pinus*, poplar *Populus* and acacia *Acacia* monocultures, rather poor in associated species.

DISTRIBUTION AND STATUS OF DRY GRASSLAND BIRDS

The distribution of typical *puszta* bird species is concentrated in the main in the southeast of the country where steppe-like habitat is found (Figure 1). The special climate (Bacsó 1961), edaphic conditions (Stefanovits 1975) and human influence (such as river regulation, deforestation, grazing of cattle) all contributed to the formation of this secondary habitat

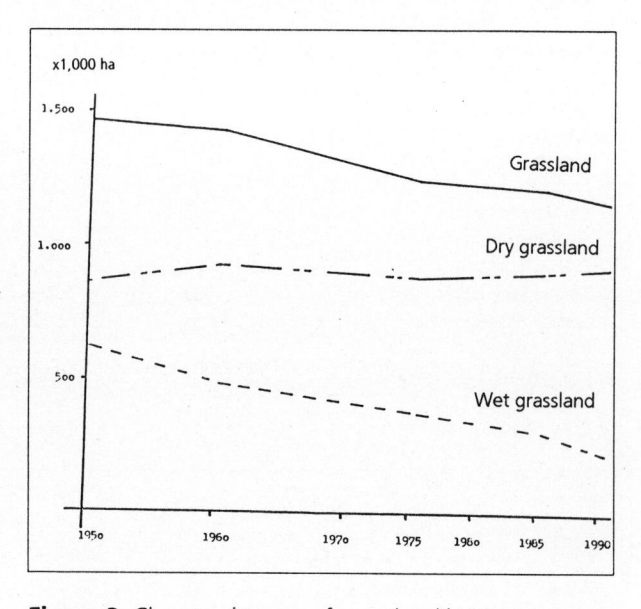

Figure 2: Changes in area of grassland in Hungary over the last 40 years

(Hortobágy and Simon 1981). If we take the distribution of the great bustard *Otis tarda* as an example, we find that it overlaps almost exactly with *puszta* habitat in Hungary (Figure 3). Some species, such as short-toed lark *Calandrella brachydactyla* (Figure 4) and collared pratincole *Glareola pratincola* occur in very few areas (Hortobágy and Kardoskút). Other species such as white stork *Ciconia ciconia* and saker falcon *Falco cherrug* occur on a nationwide basis regardless of whether or not there is *puszta*. The most recent information on Hungarian breeding bird distribution and population was collected more than eight years ago (Haraszthy 1984) and that of wintering and passage species was only summarised four years ago (Haraszthy 1988). Information on the changes in population and distribution of typical *puszta* bird species is rather fragmented. No one has dealt solely with *puszta* species as a clearly defined group. Figures 5 to 8 show the distribution of certain dry grassland birds.

The Hungarian Red Data Book (Rakonczay 1990), contains 16 species out of those listed by Goriup and Batten (1990). These species are categorized as follows:

Extinct and vanished: crane, little bustard;

Endangered: saker falcon, great bustard, collared pratincole, black-winged pratincole, short-toed lark;

Vulnerable: white stork, Montagu's harrier, grey partridge, curlew, stone curlew, barn owl;

Rare: lesser kestrel, short-eared owl, rose-coloured starling.

Of the above species, the crane *Grus grus* disappeared as a breeding species early this century (Berzsenyi 1918) and now only occurs on passage, although in increasingly large numbers, with some 20-50 over-summering. The little bustard *Tetrax tetrax* has never bred in large numbers in Hungary and, indeed, the last nesting record was as far back as 1918 (Haraszthy 1988). Table 2 summarizes the Hungarian lowland dry grassland bird species status.

Table 2: Status of Hungarian lowland dry grassland birds

Project species	Migrant status	Population size	Population trend	Hungarian Red Data Book status
Vulnerable birds (29)				
Ciconia ciconia	Common breeder; migrant	4,500 pairs	–	Vulnerable
Circus cyaneus	Winter visitor		~	
Circus macrovus	Rare on passage		~	
Circus pygargus	Breeds locally; migrant	100-500 pairs	~	Vulnerable
Buteo rufinus	Regular in autumn		~	
Aquila rapax	Rare vagrant		~	
Falco cherrug	Scattered breeder; migrant	100-150 pairs	~	Endangered
Falco columbarius	Winter visitor		~	
Falco naumanni	Rare breeder; migrant	1-10 pairs	~	Rare
Anthropoides virgo	Very rare vagrant	25 peak	~	
Grus grus	Large number on passage	20,000 peak	+	Extinct
Tetrax tetrax	Rare vagrant		~	Extinct
Otis tarda	Breeds locally; resident	1,200 individuals	–	Endangered
Burhinus oedicnemus	Breeds locally; migrant	200 pairs	~	Vulnerable
Glareola pratincola	Rare breeder; migrant	30-50 pairs	–	Endangered
Glareola nordmanni	Irregularly breeding visitor	1-2 pairs	–	Endangered
Asio flammeus	Winter visitor; rare breeder	10-50 pairs	~	Rare
Melanocorypha calandra	Vagrant		~	
Calandrella brachydactyla	Breeds locally; migrant	10-100 pairs	~	Endangered
Anthus campestris	Common breeder; migrant		~	
Other birds (15)				
Coturnix coturnix	Scattered breeder; migrant		+	
Perdix perdix	Scattered resident	30,000 individuals	–	Vulnerable
Vanellus vanellus	Common breeder; migrant		–	
Numenius arquata	Scattered breeder; more on passage	100 pairs	–	Vulnerable
Tyto alba	Scattered breeder; resident	100-1,500 pairs	–	Vulnerable
Athene noctua	Common resident	100-1,500 pairs	–	
Alauda arvensis	Common breeder; migrant		~	
Galerida cristata	Common resident		~	
Anthus pratensis	Common on passage; some winter		~	
Motocilla flava	Common breeder; migrant		–	
Oenanthe oenanthe	Common breeder; migrant		~	
Miliaria calandra	Common breeder; some winter		–	
Sturnus roseus	Rare vagrant		~	Rare

CURRENT MEASURES FOR THE CONSERVATION OF LOWLAND DRY GRASSLAND BIRDS

Site protection and management of grassland bird habitats

The protection of lowland grasslands began in the early 1970s when the nature conservation authorities prepared a long-term plan for the conservation of Hungary's *pusztas*. It was at that time too that the Hortobágy (1973) and Kiskunság (1975) National Parks were created and other extensive grassland areas put under protection (Table 1). The main reason for the above measures was the preservation of dry grasslands. However, due to a later focus of attention on forests, measures relating to grasslands proceeded only slowly (1980s). This development can be criticised as it neglected those areas which are most typical and of most natural value as natural habitat in Hungary, that is, grasslands.

Hungary covers 9,303,183 ha, of which 588,622 ha (6.3 per cent) are protected areas. Grasslands cover 1,185,570 of the country with 121,413 ha protected (equivalent to only 24 per cent of all protected areas, Figure 9). The preliminary inventory (Table 1) recommends that at least another 15,000 ha of grassland should be protected. Furthermore, there are another 250,000 ha of grassland in Hungary which are used extensively for agriculture, but which have importance from a nature conservation point of view. We need to identify possible conservation measures for the latter such as alternative agricultural methods, environmentally-friendly agricultural policies and general landscape protection laws.

On the positive side, Hungary has signed several international nature conservation conventions such as Ramsar, Bonn and Bern and the UNESCO's Biosphere programme. These international obligations, of course, cover grasslands and their bird species. There are 13 Ramsar sites in Hungary, totalling 110,389 ha. Grasslands totalling 28,226 ha automatically come under protection in six of these areas:

Hortobágy National Park	15,000 ha
Pusztaszer LPA	5,000 ha
Kiskunság National Park	3,903 ha
Lake Fertó LPA	2,870 ha
Velence-Dinnyés NCA	965 ha
Kardoskut NCA	488 ha

There are also three Biosphere Reserves in the country which are essentially dry lowland grasslands:

Hortobágy NP 52,000 ha (core area 1,285 ha)
Kiskunság NP 30,628 ha (core area 2,280 ha)
Lake Fertó LPA 12,543 ha (core area 375 ha)
Total 95,171 ha (core area 3,940 ha)

The core areas of these are protected by the strictest regulations.

Lake Fertó (Neusiedlersee) has been proposed as an international National Park.

Only one third of the total of 170 protected areas in Hungary have long-term conservation management plans. Concerning lowland grassland reserves, however, around half have long-term management plans. One of the most serious grassland management problems is under-grazing: there are simply insufficient number of appropriate livestock in Hungary. The nature conservation authorities wish to take into ownership the most important protected grasslands and manage them accordingly (Plates 2 and 3).

At present, 15,000 ha are managed by the mentioned authorities. Indeed, at this very moment the authorities are negotiating the ownership of grassland areas formerly used by the Soviet military. In the last five years the following *puszta* areas underwent habitat reconstruction: Hortobágy, Kiskunság, Lake Fertó and Kardoskut. The main aim of this work was to recreate grassland and wetland bird habitats.

Protection of species

Altogether 350 bird species have occurred in Hungary: 327 are protected and of these 38 are strictly protected. All the species mentioned by Goriup and Batten (1990) which have occurred in Hungary are protected. Table 3 presents further figures relating to Hungary's European lowland dry grassland birds.

CURRENT CONSERVATION PROJECTS

General programmes

1   European Community
PHARE-Environment Programme Wetlands and Grasslands Protection Study in Hortobágy NP and Kiskunság NP. The general objectives served by implementing the measures defined by the study would be the conservation of unique natural resources, provision of an economic base for the population in the rural areas surrounding the national parks, exploitation of the tourism potential on a sustained basis and furthering of the scientific knowledge of environmental management of these areas.

2   IUCN East European Programme
The Environmental Impacts of Intensive Agriculture in Hungary and Poland. Guidelines and training for sustainable agricultural development in Hungary and Poland through the application of western European experience of the environmental consequences of intensive agriculture.

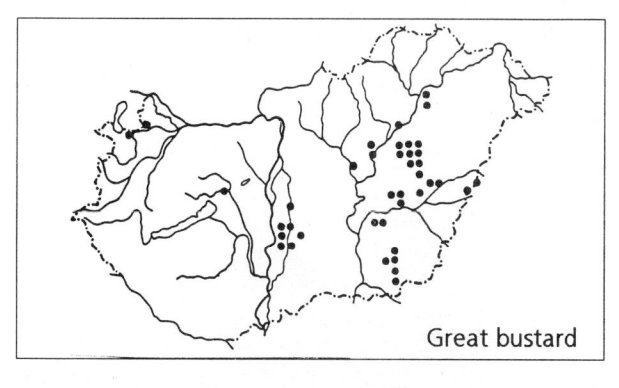

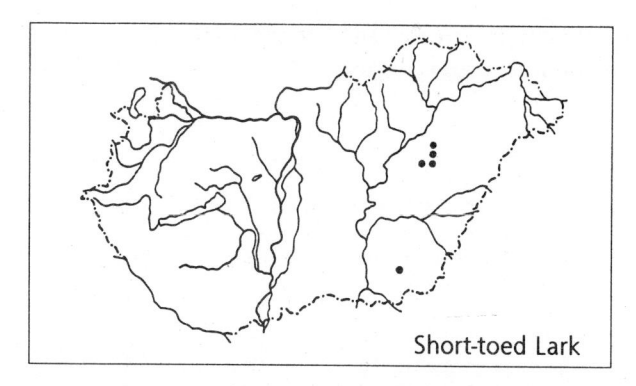

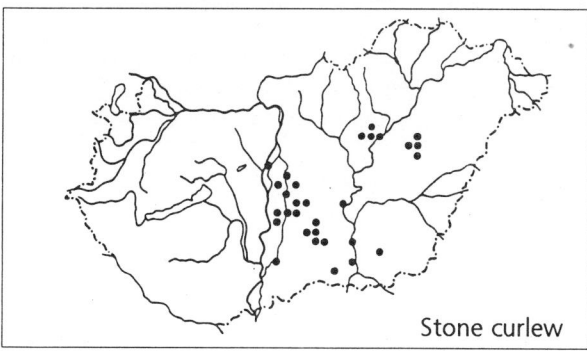

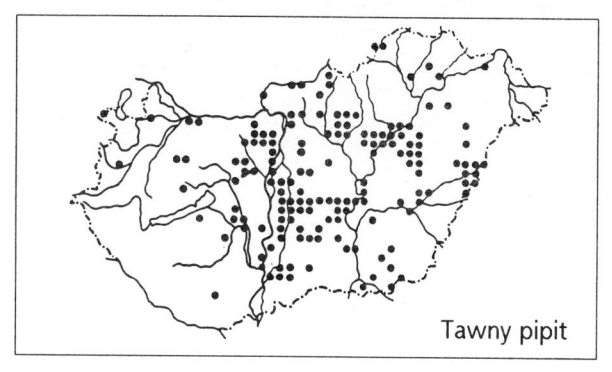

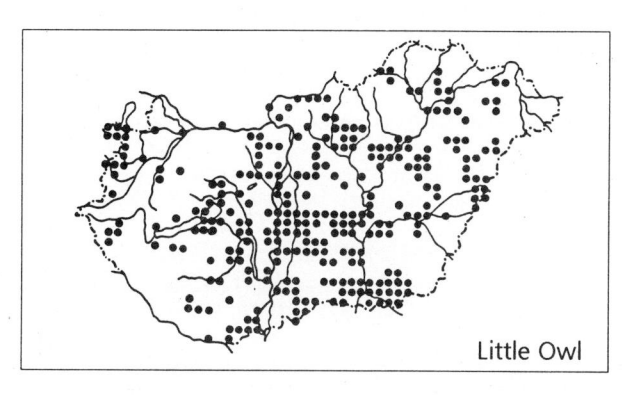

Figures 3-7 Distribution maps of selected bird species in Hungary

3   Hungarian Ministry of Environment (ME)
Environment policy projects include development of environmentally friendly agriculture.

4   ICBP Important Bird Areas Follow-up
Carried out by the Hungary/Hungarian Ornithological and Nature Conservation Society (HOS); this work monitors the sites listed in Grimmett and Jones (1989).

5   WWF - ME - HOS
Projects for the conservation of Hungarian lowland grasslands include the following:
- ME - University of Sopron
  Monitoring the status of great bustard populations
- ME - HOS
  Study on great bustard-friendly agriculture
- ME - HOS
  Study on the protection of great bustard nesting sites
- ME - KNP
  Management of threatened great bustard nesting sites in Dévaványa region
- ME - HOS
  Monitoring the status of the white stork and management of threatened nesting sites
- HOS
  Study of hen harrier roosting sites in winter
- HOS
  Management of threatened Montagu's harrier nesting sites
- WWF - HOS
  Conservation of the saker falcon
- HOS
  Management of threatened barn owl nesting sites

CONCLUSION

In Hungary the basic information needed for bird protection on lowland dry grasslands does not exist. An estimated grassland area of some 400,000 ha support valuable bird habitats. *Pusztas* of major significance are under national and international

Table 3: European dry grassland birds (see Goriup and Batten 1990) in Hungary

Number of Hungarian bird species	350
Species protected in Hungary	327
Strictly protected species	38
European dry grassland birds species	44
Number of species in Hungary	33
Vulnerable dry grassland birds in Europe	29
Vulnerable dry grassland birds in Hungary	20
Species strictly protected in Hungary	6
Other dry grassland birds in Europe	15
Other dry grassland birds in Hungary	13
Species strictly protected in Hungary	1

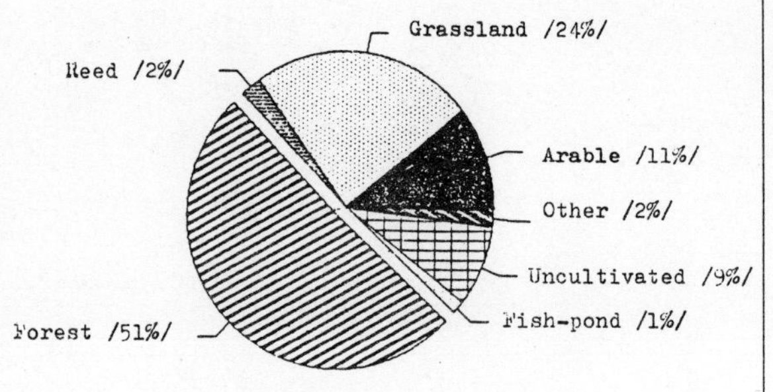

Figure 9: Land-use pattern in protected areas

protection (covering 120,000 ha) and are the subject of long-term management plans. Out of the 44 bird species listed by Goriup and Batten (1990), 33 are found in Hungary and given protected status. In recent years several programmes and actions have been launched to protect lowland habitats and their birds. However, it is necessary that more effort is made, as is clearly indicated by the decrease of the Hungarian great bustard population. What is the reason? Principally intensive agriculture development and reserve-orientated nature conservation.

Adopting a new concept for the conservation of lowland dry grasslands is essential. What are the key-factors of these further steps? Adoption of landscape conservation and rural planning law, environmentally-friendly agriculture, and an Environmentally Sensitive Area System. The scientific background and provision of experts is absolutely vital for the implementation of these requirements.

ACKNOWLEDGEMENTS

I should like to express my thanks to the following for helping with this paper: Dr I. Sterbetz, Dr J. Gyory, Dr G. Kovács, Dr Zs. Kalotás.

REFERENCES

Bacsó, N. (1961). A pusztai, atlanti és mediterrán éghajlati jellegek hatása hazánk mezogazdaságára. *Agrártudományi Egyetem Mezogazdaságtudományi Karának Közleményei 1961*: 157-172.

Berzsenyi, Z. (1918). A daru fészkelése a balatoni berekben. *Természettudományi Közlöny* 50: 124-125.

Bodrogközy, Gy. (1970). Ecology of the halophilic vegetation of the Pannonicum. *Acta Biological Szeged* 16: 21-41.

Enyedi, Gy. (1964). *A Délkelet-Alföld mezogazdasági földrajza.* [Agricultural geography of southeast lowland]. Akadémiai Kiadó, Budapest.

Goriup, P.D. and Batten, L. (1990). The conservation of steppic birds: a European perspective. *Oryx* 24: 215-223.

Grimmett, R.F.A. and Jones, T.A. (1989). *Important bird areas in Europe.* ICBP Technical Publication No. 9. ICBP, Cambridge.

Haraszthy, L. (1984). *Magyarország fészkelo madarai.* Natura, Budapest.

Haraszthy, L. (1988). *Magyarország madárvendégei.* Natura, Budapest.

Hortobágyi, T. and Simon, T. (1981). Növényföldrajz, társulástan és ökológia. Tankönyvkiadó, Budapest.

Rakonczay, Z. (1990). *Vörös Könyv.* [Red Data Book]. Akadémiai Kiadó, Budapest.

Stefanovits, P. (1975). Talajtan. Mezogazdasági Kiadó, Budapest.

Sterbetz, I. (1975). Development of the nesting ornithofauna on biotopes used as pasture and hayfield in eastern Hungarian steppes. *Debreceni Déri Múzeum Évkönyvébol 1974*: 157-170.

THE GREAT BUSTARD IN THE USSR: STATUS AND CONSERVATION

Vladimir E. Flint and Alexandr L. Mishchenko[*]

All-Union Research Institute of Nature Conservation and Reserves, USSR State Committee for Environmental Protection, Znamenskoye-Sadki, PO Vilar, Moscow, USSR

INTRODUCTION

Up to the beginning of the 20th century, the great bustard *Otis tarda* nested across the entire steppe zone of the USSR, from the country's western borders to Transbaykal, and also penetrated locally into the wooded steppe and semi-desert zones. In the following decades, uncontrolled hunting and the ploughing-up of steppelands led to a decline in numbers and range contraction, though in some areas the great bustard managed to adapt to the new agricultural habitat. The situation worsened catastrophically in the 1950s and 1960s, during the campaign to plough up the virgin lands which was accompanied by an increase in uncontrolled hunting and unrestricted use of especially toxic pesticides. bustard numbers started to fall sharply and the formerly extensive range became highly fragmented.

PROTECTION AND STATUS

The great bustard is included in the Red Data Book of the USSR and Red Books of certain Union republics. Its total population in the USSR is at present not more than 8,000 birds, and some 6,000-7,000 of these are now confined to the Saratov region and adjoining districts of the Volgograd region (Lower Volga). This population is the most stable and has to a significant extent adapted to the agricultural regime operating in the area. There now appear to be signs of a slow increase in the Lower Volga population, but this has to be viewed with caution as it may merely reflect the fact that improved census methods have produced more comprehensive and accurate results.

During 1984-1990, we carried out studies in an area of 33,000 ha of the 44,000 ha special Saratov sanctuary which was set up to conserve the great bustard. The annual breeding population in the study area numbered 65-90 females. The area is typical of the steppe-lands in that part of Saratov region which lies east of the Volga river: almost all the land is under the plough and most fields are given over to the cultivation of cereal crops (wheat, millet, barley, rye, maize, etc.). In both the study area and Saratov region as a whole, great bustards nest almost without exception on fields which at the time of laying are either bare plough or have emerging cereal shoots (up to 15 cm). Hens show a preference for slightly undulating fields with an uneven, mosaic micro-relief. Mean clutch size in 1984-1990 was 1.93 eggs (n=155), and peak laying time is the first third of May.

Nest destruction and predation

Agricultural activity either finishes before the bustards start nesting or begins after hatching on only 15-20% of the fields. On other fields, up to 90-95% of nests are destroyed by agricultural work (especially when this is carried out at night). In the study area, our data show that only 20-25% of first clutches survive. This percentage increases as replacement clutches are laid (after the loss of first clutches) up to the beginning of June. These figures are evidently also valid for Saratov region as a whole.

Nests are lost on fields either directly through crushing by farm machinery (on fields where the cereal crop is well advanced or where work continues at night), or the eggs are destroyed by rooks *Corvus frugilegus*. Flocks of rooks regularly follow farm tractors and when a hen bustard is disturbed and leaves the nest, they immediately move in and peck open the eggs. Since the 1960s, there has been a a very large increase in the number of rooks nesting in Saratov and adjoining regions, the main reasons for this being as follows: trees in field shelter-belts (which were planted in the 1930s and 1940s to help keep

[*]Translated from the Russian text by Mike Wilson, Editor, *Birds of the Western Palearctic*

snow on the fields) had grown tall enough for rooks to nest in them; natural predators (eagle owl *Bubo bubo* and saker *Falco cherrug*) had disappeared; and there had been a considerable improvement and stabilization of the food-base. Rookeries containing 1,000-2,000 pairs are now not uncommon, and effective measures to control their numbers have yet to be discovered or put into practice. The fox population is relatively small and predation of great bustards by foxes is insignificant.

There are as yet practically no data on the effect of pesticides on great bustards. Our study has shown that on fields treated with pesticides containing mercury and chlorine applied by tractor-drawn spray all hen bustards desert their nests.

Conservation measures

Since 1982, a special programme, designed to protect the great bustard and to increase its numbers, has been in operation in Saratov region. This programme includes protection for bustard display, nesting and feeding grounds, seeks to ensure that nature protection laws are strictly observed and that the conservation message is widely disseminated; the programme further includes the establishment of a captive breeding population in a special bustard farm and the eventual release of captive-bred birds, both to bolster existing wild stock and to re-establish wild populations in areas where the species has become extinct. Preserving great bustard habitat will also benefit other rare grassland species such as little bustard *Tetrax tetrax*, demoiselle crane *Anthropoides virgo*, steppe eagle *Aquila rapax*, collared pratincole *Glareola pratincola*, and quail *Coturnix coturnix*.

In 1984-1988, drawing on the benefit of experience gained in similar programmes outside the Soviet Union, methods were perfected for artificial incubation of eggs and captive rearing of chicks, including development of the correct diet for young bustards. As the work was carried out in small and temporary stations which were inadequately staffed and equipped, considerable difficulties arose when it came to getting the bustards through the winter, and many young birds unfortunately perished. In 1991, a new and well-equipped bustard breeding station is due to come into operation in Engels (which lies on the east bank of the Volga, opposite Saratov); the station will have facilities for artificial incubation of 400 eggs and the captive stock will comprise at least 200 adult great bustards. The aim is the successful implementation of a great bustard conservation programme centred on the new captive-breeding station, and it is therefore planned to work with artificial-insemination techniques and to ensure that young bustards are reared having no contact with humans in order to prepare them for release into the wild; suitable methods also need to be developed for releasing the bustards at various sites in Russia and the Ukraine (release on wintering grounds and in spring at display sites).

CONCLUSION

The development of international co-operation is of paramount importance. Young bustards from the Saratov population could be a good basis both for bolstering declining populations in Europe (e.g. Germany, Austria, Czechoslovakia), and for re-introduction in countries where the species has become extinct (e.g. Britain).

Plate 1: Great bustards on arable farmland in the Saratov region [photo: V E Flint]

CONSERVATION PROBLEMS OF STEPPIC AVIFAUNA IN TURKEY

Y. Sancar Baris

Ondokuzmayis Universitesi Tip Fakultesi, Patoloji Ana Bilim Dali, 55139 Kurupelit, Samsun, Turkey

ABSTRACT

Environmental conservation is a relatively new concept in Turkey and steppic habitats have been greatly neglected. A substantial part of the Turkish landscape is covered by steppes and they mostly suffer from over-grazing. Several endangered bird species breed in Turkish steppes but their distributions and populations are poorly known. It seems that steppic habitats are not as seriously threatened in Turkey as they are elsewhere in Europe but immediate measures as well as long term research and education projects must be realized to ensure the future existence of steppic habitats and their birds.

INTRODUCTION

Due to its distinctive geographical position Turkish fauna and flora is highly diverse, the vegetation has been influenced by Euro-Siberian, Mediterranean and Irano-Turanian phytogeographical regions (Ertan *et al.* 1989). Main topographical features (Figure 1) are mountains and plateaux extensively covered by steppe. Despite their dominance in the landscape, steppic habitats have been largely neglected as the limited ornithological research and conservation activities have concentrated on relatively better known areas like wetlands; thus, the ecology of the Turkish grasslands and their problems remain largely unknown. This paper aims to present the general features of the steppic habitats and the related avifauna in Turkey, with an attempt to emphasize conservation problems and priorities.

DESCRIPTION AND DISTRIBUTION OF STEPPIC HABITATS IN TURKEY

The main range of steppic habitats in Turkey extends over the whole of central, east and southeast Anatolia (Figure 2). The term steppic habitats is taken here to include the primary or natural steppe, areas covered by a steppe-like secondary vegetation often formed by clearing and grazing of a former scrub or forest area and agricultural areas with low-intensity cereal and leguminous crops (Goriup 1988). Natural steppe, originally confined to the plains surrounding Tuz Gölü in central Anatolia, is characterized by *Artemisia* and *Thymus* species often in association with chenopods (Chenopodiaceae) like *Noaea* species (Ozenda 1979). A secondary steppe with similar vegetation covers the rest of central and east Anatolia with remnant islands of oak *Quercus*, pine *Pinus* and juniper *Juniperus* (Ozenda 1979). In higher altitudes and/or in over-grazed areas more resistant and less palatable plants like *Astralagus* and *Achantholimon* dominate the flora (Kence 1987).

The steppes of southeast Anatolia are of a semi-desert character and some of the typical plants are *Artemisia herba-alba*, *Eryngium noerium* and *Salvia spinosa* (Ertan *et al.* 1989). In addition to this main range, degradation of the mesomediterranean and thermomediterranean vegetation in the Mediterranean, Aegean and Marmara regions has led to the formation of a steppe-like vegetation with *Bromus* and *Medicago* species, especially productive in winter and drying in summer (Tukel 1988). Finally, saline lakes and deltas support a halophytic vegetation often with *Salicornia* and *Salsola* species (Ertan *et al.* 1989, Nurhat and Ortac 1984). Grasslands of eastern Anatolia and Black Sea regions have an average yearly production of 820 kg/ha and 50-60 per cent vegetation coverage whereas those in central and southeast Anatolia rank the least productive with an average yearly production of 300/kg/ha and only 10-15 per cent coverage (Genckan *et al.* 1990). The facts that the former regions receive 500-1000 mm annual rainfall compared with 300 mm/year (Nurhat and Ortac 1984) and a higher population density in the latter form the possible basis for this difference.

Determining the range of the agricultural pseudo-steppe is problematic because dry agriculture with fallow, often seeded in rotation, account for 60.6 per cent of all the agricultural land in Turkey (Oztan *et al.* 1986). The most important crops are wheat *Triticum*, barley *Hordeum*, oat *Avena*, and (locally) maize *Zea*. Various pulses and fodder grasses are cultivated mainly in the eastern half of the country. Table 1 shows the area covered by the major land use groups. Grassland alone cover 21.7 million hectares, 28 per cent of the total area. Steppic habitats cover a total of 43.7 million hectares and extend over about half of the country's surface area.

STEPPIC AVIFAUNA

The Turkish avifauna is highly diverse, with 371 species from 180 genera known to occur regularly, including 186 breeding species (Baris 1989). The steppic zone depicted in Figure 2 is inhabited by 96 breeding species from 65 genera (Table 2). Of these, 42 species (44 per cent) are ground-nesters, best represented by larks (Alaudidae, nine species) and gamebirds (Phasianidae, five species). Bustards (*Otis tarda*, Otididae), stone curlews (*Burhinus oedicnemus*, Burhinidae) and sandgrouse (Pteroclididae) are entirely confined to steppic habitats (Goriup 1988). Gamebirds and larks are highly characteristic of steppic habitats but their ranges are not confined to the steppic zone. Other bird species do not nest on the ground but depend largely on steppic habitats for hunting and foraging. This group is best represented by diurnal birds of prey (Accipitridae and Falconidae), with a total of 15 species. Members of the families Glareolidae, pratincoles and Charadriidae (waders) inhabit saline grasslands and agricultural areas surrounding open water. Finally, mention should be made of species like white stork *Ciconia ciconia* and crane *Grus grus* which are definitely not steppe-breeders but use steppic habitats extensively during migration and/or wintering.

CONSERVATION PROBLEMS

Grazing

In Turkey, over-grazing is perhaps the single most important problem in the conservation of the primary and secondary steppe. Standardized to cattle head, the Turkish national herd numbered 28.6 million BHU in 1985 compared to 21.0 million BHU in 1950 (Oztan *et al.* 1986. [BHU: Bovine Head Unit; one BHU = a 250 kg cow or equivalent, modified for Turkey (Tukel 1988)]. During the same period, the grassland area considerably diminished from 37.8 million hectares in 1950 to 21.7 million hectares in 1985; thus the grassland area per BHU was 1.79 ha/BHU in 1950 and 0.86 ha/BHU in 1985 (Oztan *et al.* 1986). Given the fact that 68.8 per cent of the national herd's subsistence depends on grasslands (Tosun 1977, in Tukel 1988), one could safely state that during the 35 year period between 1950 and 1985, grazing pressure on the Turkish grasslands has at least doubled.

Over-grazing can be even more serious on a regional scale, especially in the less productive grasslands of central and southeast Anatolia. Indeed, demand for more and better grazing area is so intense that often nearby scrub and forest is seriously damaged (Oztan *et al.* 1986). Carrying capacities and actual grazing pressures in different geographical regions are presented in Table 3. The issue is further complicated by other forms of bad grazing management such as prolonged grazing and grazing in early spring, during the seed formation period and in late autumn, the critical periods of the plant life cycle (Tukel 1988). As a result, over-grazing reduces the productivity and plant cover of the grasslands and inevitably affects the species composition of the flora, destroying the climax species and selecting less palatable ones.

Destruction by agriculture

In 1950, 59.8 per cent of all the land was covered by grassland (steppes, highland grassland and moorland) whereas they covered only 31.1 per cent of the land in 1984; during the same period agricultural land coverage increased from 20.6 per cent to 34.1 per cent (Tosun and Altun 1986).

The majority of the land reclaimed from steppe is of a pseudo-steppe character with non-irrigated cereal and pulse cultivation and fallow. In fact, dry agriculture with fallow accounts for 60.6 per cent of all the agricultural land in Turkey (Oztan *et al.* 1986) and although the demand for new agricultural areas is considerable, it seems that in general the habitat destruction by agriculture should not be considered an immediate threat unless widespread agricultural intensification occurs. However, immense irrigation projects presently under construction in southeast Anatolia show that such a radical change in the agricultural policy is just about to happen. With the realisation of this project, 1.7 million hectares of steppe and semi-desert land will be opened to irrigation and intensive farming (Ozbay 1990).

The area concerned has a very distinctive fauna and species like see-see partridge *Ammoperdix griseogularis* and pin-tailed sandgrouse *Pterocles alchata* have their distributions in Turkey limited to the region. Furthermore the area might prove to support an important population of breeding and wintering great bustards *Otis tarda* (Goriup 1981, Martins 1989). The project's impact on the local habitats, fauna and flora awaits further investigation. It must also be noted that the total increase in area covered by agriculture during 1950-1984 accounts for only half of the steppe area lost, therefore during the same period an equal amount of steppe must have been reforested, urbanized or dammed. The habitats created by these activities are completely different

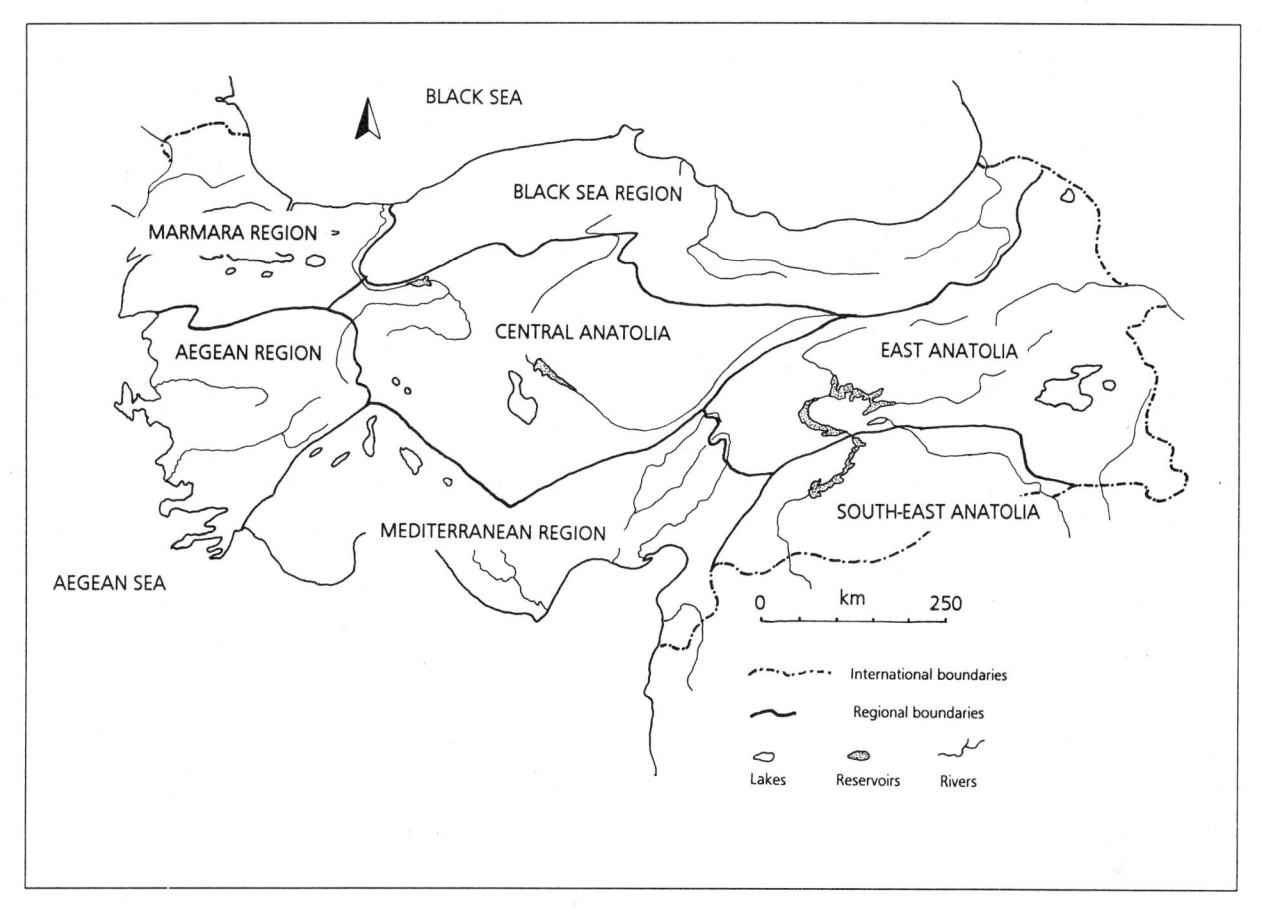

Figure 1: Geographical regions of Turkey (after Erol 1982)

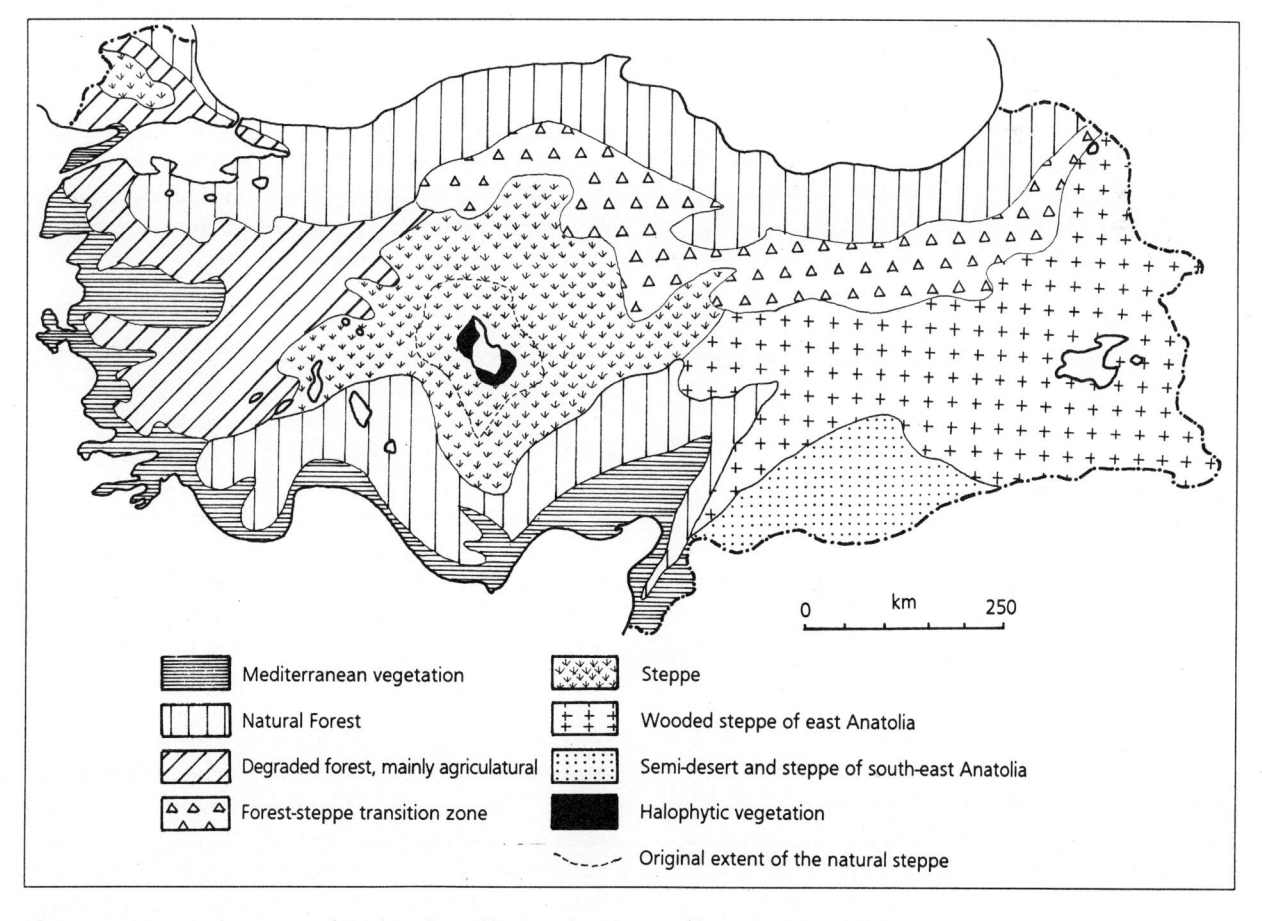

Figure 2: Vegetation map of Turkey (modified and redrawn after Ozenda 1979)

from steppic habitats and thus although the area destroyed by each is smaller, the effect is much more deleterious.

Endangered species

In Turkey, the lack of reliable and comparable ornithological data makes it difficult to estimate numbers and determine population trends for most species. Key species to monitor any specific habitat have never been defined and there is no Turkish red data book to present the guidelines for future conservation work. Goriup and Batten (1990) have listed 44 key dry grassland species breeding in Europe (taken here to include all of Turkey), 25 of these are considered vulnerable (Grimmett and Jones 1989).

Those species which breed regularly in Turkey are presented in Table 4; a few other species like chukar *Alectoris chukar*, black francolin *Francolinus francolinus* and see-see partridge might be included in a purely Turkish list. Some amongst the vulnerable species are endangered on a global scale: lesser kestrel *Falco naumanni*, demoiselle crane *Anthropoides virgo* and great bustard *Otis tarda* (Grimmett and Jones 1989) and deserve a brief discussion of their present status in Turkey.

The lesser kestrel is a fairly widespread and locally common summer visitor (Cramp and Simmons 1977) but the population's size, trend, and detailed distribution are not known. The demoiselle crane is a very local summer visitor to breeding grounds in east Anatolia, with an estimated population of 20-30 pairs apparently under no immediate threat (Kasparek 1988). The great bustard is a local resident and partial migrant in central, east and southeast Anatolia (Martins 1989). There has been some decline in the past but the survey by Goriup and Parr (1981) suggest that the species is more numerous than previous data suggested and that Turkey might be one of the most important centres of distribution (Goriup and Parr 1981). It is also worth mentioning that the status of little bustard *Tetrax tetrax*, which has bred in southeast Anatolia in the past, is at present uncertain and that this species might still breed in small numbers in central Anatolia (Martins 1989). The turkish breeding range and/or distributions of other vulnerable species in Table 4 might be of a regional importance since, although somewhat speculative, they may account for a significant part of the European range and/or populations. Long-legged buzzard *Buteo rufinus*, crane *Grus grus*, collared pratincole *Glareola pratincola*, greater sand plover *Charadrius leschenaultii*, black-bellied *Pterocles orientalis* and pin-tailed sandgrouse *P. alchata* and lesser short-toed lark *Calandrella rufescens* are the most important of this group; indeed, the European breeding range of greater sand plover is almost entirely confined to Turkey.

Table 2: Steppe breeding birds in Turkey

Families	Species	No. ground nesting
Ciconiidae	1	-
Accipitridae	9	1
Falconidae	6	-
Phasianidae	5	5
Gruidae	1	1
Otididae	1	1
Burhinidae	1	1
Glareolidae	1	1
Charadridae	4	4
Pteroclidae	2	2
Columbidae	3	-
Cuculidae	2	-
Tytonidae	1	-
Stringidae	2	-
Caprimulgidae	1	1
Apodidae	1	-
Meropidae	2	2
Coracidae	1	1
Upupidae	1	-
Alaudidae	9	7
Hirundidae	3	-
Motacillidae	3	2
Pycnonotidae	1	-
Turdidae	9	5
Sylvidae	4	-
Laniidae	3	-
Corvidae	4	-
Ploceidae	3	-
Fringillidae	5	2
Emberizidae	6	4

Grassland legislation in Turkey

At present, legal provisions are mainly concerned with the regulation of the usage of a grassland by its human inhabitants and not with the protection of a defined habitat. All grasslands are state-owned but are loaned for common public use; accordingly grasslands cannot be traded, their "character" cannot be changed and their range cannot be extended (Tosun and Altun 1986). The law thus bans the transformation of grassland to agricultural land and any other activity which may result in the loss of grassland. However, the definition of the grazing areas is somewhat blurred and with no governmental organisation directly responsible for grasslands, the law is poorly enforced.

Present legislation also does not provide regulations to control stock levels and grazing periods (Bakir and Erkun 1990). Obviously there is an

Table 1: Land-use groups and their coverage (Oztan *et al.* 1986)

Land-use groups	Area (million ha)	%
Dry agriculture without fallow	5.8	7.5
Dry agriculture with fallow	16.8	21.6
Irrigated fields	3.0	3.8
Orchards, vegetable gardens etc.	1.1	1.4
Special crops (tobacco, tea, etc.)	1.0	1.3
Sub-total agriculture	**27.7**	**35.6**
Moist pastures	0.6	0.8
Dry grasslands	21.1	27.1
Sub-total grasslands	**21.7**	**27.9**
Forest and scrub	23.5	30.2
Urban areas	0.6	0.8
Other areas (dunes, open water surfaces etc)	4.3	5.5
TOTAL	**77.8**	**100.0**

Table 3: Grazing pressure in different geographical regions (Oztan et al. 1986).

Geographical region	Grassland area (million ha)	Herd size (million BHU)	Grazing pressure (Ha/BHU)	Carrying capacity (Ha/BHU)
Black Sea	1.7	5.8	0.3	1.2
Marmara	0.5	3.3	0.2	1.8
Aegean	1.0	3.2	0.3	1.8
Mediterranean	0.7	2.6	0.3	2.5
Central Anatolia	6.2	6.6	0.9	2.7
East Anatolia	8.4	5.0	1.7	1.0
South-east Anatolia	3.3	2.2	1.5	3.2

immediate need for a revision of the present grassland legislation which should determine realistic regulations to control the grazing pressure, define general grassland management measures and assign a government organisation responsible for its enforcement.

Hunting

All vulnerable species are legally protected throughout the year with the exception of sandgrouse for which hunting is allowed between 15th August and 28th February (Central Hunting Committee decisions for the 1990-1991 hunting season). All gamebirds except grey partridge and quail are fully protected. Lapwing *Vanellus vanellus* and curlew *Numenius arquata* are protected during the breeding season and only crows may be shot throughout the year. Hunting legislation in fact provides protection for most species but it must be kept in mind that hunting laws are poorly enforced and often severely violated in Turkey (Magnin 1989). Nevertheless, there seems to be no severe hunting pressure on most of the endangered species and recent works suggest that great bustard hunting is possibly of no significant importance, contrary to previous thinking (Goriup and Parr 1981).

Protected areas

Turkey has no nature reserves established specially to protect steppic fauna and flora. Some of the protected important bird areas do have steppic habitats within their boundaries (Ertan et al. 1989) but none of them have a substantial part of their area covered by steppes (Goriup 1988). On the other hand, Goriup and Parr (1981) point out the effective protection provided by remote state-farms in Turkey and their possible importance in the conservation of the great bustard.

CONSERVATION PRIORITIES

For obvious reasons, immediate measures to protect the endangered species have to be taken, and these should include establishing grassland reserves for demoiselle crane and great bustard. For the latter species, significant improvements can be gained simply by education of and co-operation with the state-farm staff and the relevant central government authorities. Other urgent actions should include surveys and short term projects to determine the populations and distributions of lesser kestrel, great bustard, greater sand plover and pin-tailed sandgrouse. Immediate research is also needed to determine the possible environmental effects of large-scale irrigation projects. Medium to long term plans should be made for the preparation of appropriate grassland legislation and although the responsibility here clearly belongs to the Turkish State, national and international organisations should be prepared and motivated to introduce the law-

Table 4: Vulnerable dry grassland birds breeding regularly in Turkey

White stork	*Ciconia ciconia*
Montagu's harrier	*Circus pygargus*
Long-legged buzzard	*Buteo rufinus*
Saker falcon	*Falco cherrug*
Lanner falcon	*Falco biarmicus*
Lesser kestrel	*Falco naumanni*
Demoiselle crane	*Anthropoides virgo*
Crane	*Grus grus*
Great bustard	*Otis tarda*
Stone-curlew	*Burhinus oedicnemus*
Collared pratincole	*Glareola pratincola*
Greater sand plover	*Charadrius leschenaultii*
Black-bellied sandgrouse	*Pterocles orientalis*
Pin-tailed sandgrouse	*Pterocles alchata*
Calandra lark	*Melanocorypha calandra*
Short-toed lark	*Calandrella brachydactyla*
Lesser short-toed lark	*Calandrella rufescens*
Tawny pipit	*Anthus campestris*

maker to the concept of grassland protection. Finally, long to very long term plans should be prepared for public education and a widespread and efficient national ornithological coverage.

CONCLUSION

Turkish steppes with their extensive range, varied character and avifauna, deserve attention and may be very important for the conservation of steppic habitats and steppic avifauna in Europe and the Middle East as a whole. Although threatened by overgrazing as well as other factors, Turkish steppes seem to be in a better condition in terms of both quality and quantity compared with many other European countries. Immediate conservation measures as well as long term education and research projects must be undertaken to preserve the steppic flora and its birds. A significant improvement in this respect could be accomplished with the preparation and enforcement of a realistic grassland conservation law.

ACKNOWLEDGEMENTS

I thank Sabahat Ozman for her help in providing basic material for the text and to BEP Bilgisayar for providing word-processing facilities.

REFERENCES

Anon. (1990). *1990-91 av mevsimi Merkez Av Komisyonu kararlari* Orman Genel Mudurlugu, Tarim Orman ve Koyisleri Bakanligi.Ankara.

Bakir, O. and Erkun, V. (1990). Mer'a kanunu. *Tarim Orman ve Koyisleri Bakanligi Dergisi*. 51:11-13.

Baris, Y. S. (1989). Turkey's bird habitats and ornithological importance. *Sandgrouse* 11: 42-51.

Cramp, S. and Simmons, K.E.L. (eds.) (1977). *The birds of the Western Palearctic*. Vol.1 Oxford University Press.

Erol, O. (1982). Turkei: Naturraumliche Gliederung. *Tübinger Atlas des Vorderen Orients*. Wiesbaden.

Ertan, A., Kilic, A. and Kasparek, M. (1989). *Turkiye'nin onemli kus alanlari*. DHKD, Istanbul.

Genckan, M. S., Avcioglu, R. and Okuyucu, F. (1990). Cayirmer'alarimizin durumu. *Tarim Orman ve Koyisleri Bakanligi Dergisi*. 51: 8-10.

Goriup, P.D. (1988). The avifauna and conservation of steppic habitats in Western Europe, North Africa and the Middle East. In: Goriup, P.D. (ed.), *Ecology and conservation of grassland birds*. ICBP Technical Publication No.7. ICBP, Cambridge.

Goriup, P.D. and Batten, L. (1990). The conservation of steppic birds: a European perspective. *Oryx* 24: 215-223.

Goriup, P.D. and Parr, D. (1981). Report on a survey of bustards in Turkey, 1981. *ICBP Study Report* No.1

Grimmett, R.F.A. and Jones, T.A. (1989). *Important bird areas in Europe*. ICBP Technical Publication No.9. ICBP, Cambridge.

Kasparek, M. (1988). The demoiselle crane, *Anthropoides virgo*, in Turkey: distribution and population of a highly endangered species. *Zoology of the Middle East* 2: 31-38.

Kence, A. (ed.) (1987). *Biological diversity in Turkey*. Environmental Problems Foundation of Turkey, Ankara.

Magnin, G. (1989). Falconry and hunting in Turkey during 1987. *ICBP Study Report* No.34.

Martins, R.P. (1989). Turkey bird report 1982-6. *Sandgrouse* 11: 1-41.

Nurhat, C. and Ortac, A. (eds.) (1984). *Yurt Ansiklopedisi*. Vol.11. Anadolu Yayincilik, Istanbul.

Ozbey, E. (1990). GAP uzerine dialoglar 5. *Tarim Orman ve Koyisleri Bakanligi Dergisi*. 51: 42-45.

Ozenda, P. (1979). *Vegetation map of the Council of Europe member states*. Nature and Environment Ser. 16. Strasbourg.

Oztan, Y., Bakir, O. and Ekim. T. (Eds.) (1986).*Turkiye'nin cevre sorunlari 85*. Turkiye Cevre Sorunlari Vakfi, Ankara.

Tosun, F. and Altun M. (1986). *Cayir-mer'a-yayla kulturu ve bunlardan faydalanma yontemleri*. Ondokuz Mayis Universitesi Yayinlari No.5,Samsun

Tukel, T. (1988). *Cayir-mer'a amenajmani* Cukurova Universitesi Ziraat Fakultesi ders kitabi No.17, Adana.

ACTION FOR DRY GRASSLAND BIRDS IN BRITAIN

Richard F Porter, Graham D Elliott and Gwyn Williams

Royal Society for the Protection of Birds, The Lodge, Sandy, Bedfordshire SG19 2DL, UK

INTRODUCTION

In terms of bird conservation priorities in Britain dry grassland species and the dry grassland habitat rank relatively low. In general, it is only where the grassland habitat becomes flooded or partially flooded or where it is invaded by scrub that, with few exceptions, it is important for birds (see e.g. Fuller 1982). For example, in identifying its priorities for action to conserve habitats, the Royal Society for the Protection of Birds (RSPB) has not listed dry grasslands for attention.

In compiling the list of *Red Data Birds in Britain* the Nature Conservancy Council (NCC) and RSPB identified 117 species of most concern in Britain (Batten *et al.* 1990). In addition to including those that are rare, very local or rapidly declining it also includes, unlike other Red Data Books, species of international importance because of the significant numbers — in European or World terms — that breed or winter in the British Isles. The inclusion of these internationally important species is irrespective of whether they are threatened or not. The criteria for qualification as a Red Data Species are given in Appendix 1.

The species in *Red Data Birds in Britain* which make use of dry grasslands to a greater or lesser extent are given in Table 1. None are dependent on this habitat.

In addition, the following 'candidate' Red Data species which have some dependence on dry grasslands have been identified (species which do not quite meet the criteria for inclusion in Red Data Birds in Britain but which could if conditions worsen):

Red-legged partridge *Alectoris rufa*
Lapwing *Vanellus vanellus*
Short-eared owl *Asio flammeus*
Yellow wagtail *Motacilla flava*
Wheatear *Oenanthe oenanthe*
Corn bunting *Miliaria calandra*

Red Data Birds in Britain comprises a series of species texts that summarises their conservation importance, legal status, ecology, distribution and population, threats and conservation action necessary to address these threats. This is, however, only the first step. The next is to prepare Species Action Plans for each species, and this the RSPB has now embarked on working in association with the Joint Nature Conservation Committee and, where appropriate, the Wildfowl and Wetlands Trust. Currently 20 plans are near completion.

The purpose of Species Action Plans is to ensure that the best possible action is taken to conserve important, rare and threatened birds in the UK, Channel Islands and Isle of Man. It follows that any individual plan is therefore a means of defining action for a particular species. Plans serve the following purposes:

- to be a formal mechanism for converting research findings into practical action;
- to set goals against which the success of the action can be measured;
- to structure the action into discrete parts for execution;
- to identify and assign managerial responsibilities involved.

The Contents of a Species Action Plan are in three parts:

Part 1 Summary
Conservation status, priority afforded to the species, targets to be achieved, broad policies, actions and procedures for reviewing effectiveness of actions.

Part 2 Biological Assessment
Legal status, ecology, distribution and population, limiting factors (threats) and resumé of conservation action taken to date.

Part 3 Actions and Work Programmes
These are covered under the following headings:

Policy and Legislative
Site Safeguard
Land Acquisition and Reserve Management
Species Management and Protection
Advisory
International
Future Research and Monitoring
Communication and Publicity

PROPOSED ACTIONS FOR BRITAIN'S DRY GRASSLAND BIRDS

The following brief summaries cover the main conservation actions identified for Britain's "dry grassland" species. In several of these accounts the common action of maintaining suitable habitats through EC agricultural policy mechanisms such as the establishment of Environmentally Sensitive Areas (ESAs), set-aside and extensification will be apparent.

Hen harrier
Currently restricted almost entirely to heather moorland during the breeding season, but in winter a significant proportion of the population, plus immigrants, utilise lowland grasslands and arable habitats. No specific grassland conservation actions are currently necessary, but might become so if the breeding population spreads to these areas; this is unlikely in the short to medium term. The wintering population, however, will benefit from measures to conserve grasslands. The main issues to be addressed first are illegal persecution and forestry incursion in the uplands.

Montagu's harrier
Small breeding population restricted to cereal crops. Never widespread and common, but formerly occurred in a wider range of lowland habitats including reed beds and forestry plantations. The main current issue to be addressed is the prevention of accidental destruction of nests by farm machinery. If the species could be encouraged to nest in more natural habitats accidental destruction would be less of a problem. Extensification and, particularly, set-aside are likely to be beneficial, even in the short term, as this would increase the area available for successful hunting.

Merlin
Currently restricted to the uplands, notably heather moorland, but increasingly occurring in forestry plantations, and sometimes surrounding upland grasslands. Research has shown that much feeding takes place over rough grasslands and enclosed pasture adjacent to heather moorland especially in the early part of the breeding season. The restoration of such areas through ESAs and heather regeneration schemes are the most important actions.

Table 1: "Dry grassland" Red Data Birds in Britain showing population, current trends and threats

	Population (pairs)		Main threats	Secondary threats
Hen harrier *Circus cyaneus*	550	stable	Afforestation, illegal killing	Agricultural intensification
Montagu's harrier *Circus pygargus*	10	increasing	Accidental killing (agriculture)	
Merlin *Falco columbarius*	550-650	decreasing	Agricultural intensification, afforestation	Pesticides, illegal killing
Grey partridge *Perdix perdix*	few million	decreasing	Agricultural intensification, pesticides, predation	
Corncrake *Crex crex*	550-600	decreasing	Agricultural intensification, accidental killing (agriculture)	Predation
Stone curlew *Burhinus oedicnemus*	160	stable	Agricultural intensification, accidental killing (agriculture)	Egg collecting, recreational pressure, predation
Barn owl *Tyto alba*	5,000	stable	Agricultural intensification	Pesticides
Chough *Pyrrhocorax pyrrhocorax*	250-300	stable	Agricultural intensification or abandonment	Recreational pressure
Quail *Coturnix coturnix*	70-250 calling males	fluctuates	—	—
Woodlark *Lullula arborea*	370	fluctuates	—	Agricultural improvement, built development, lack of management

Plate 1: Stone curlew [photo: Paul Goriup]

Grey partridge
Has declined dramatically in the last 50 years due to processes of intensive farming. While the non-spraying of headlands makes a useful contribution to its conservation, measures such as extensification and set-aside are considered to be most important.

Corncrake
Has declined dramatically this century due to agricultural arable intensification and loss of hay meadows. The immediate main actions are the protection of adults, nests and young from destruction through the mowing of grass or hay for silage. It is proposed that this should be achieved by locating calling males, and targeting the following actions to a 200 m radius: delay cutting of silage or hay, encourage crofters and farmers to use alternative mowing techniques (in which fields are mown in strips or in a circular manner from inside out, instead of outside in as usual, so that continuous grass cover is maintained with the edge of the field to permit corncrakes to escape), slow the speed of mowing and ask farmers and crofters to be highly vigilant when cutting and to be prepared to stop if adults or young are seen. In Northern Ireland farmers have been grant aided by DoE(NI) to delay the cutting of hay or silage until chicks are mobile. In the medium term the goals are the creation of ESAs to encourage hay production and preventing such land being used for intensive silage, permanent sheep grazing or being abandoned to rough grass or scrub. In the long term the promotion of a European Community hay premium scheme is considered important (Taylor and Dixon 1990).

Stone curlew
A large decline occurred in the last 30 years but the population is now probably stable. The bulk of the population is now on arable land and the main immediate action is to prevent accidental destruction of nests, eggs and young from farm operations. Demographic data from population monitoring shows that this is a necessary short to medium term measure to prevent the population declining further.

In the medium to long-term, the objective is to encourage the bulk of the population back onto semi-natural habitats, particularly Breckland heaths and chalk downland, where it would no longer be vulnerable to farm operations. Such measures will include promotion of ESAs and set-aside, involving: (i) improved management of existing grassland habitats by encouraging more intensive grazing with sheep and rabbits; and (ii) grassland re-creation from arable as well as restoration of grassland from scrub created by cessation of grazing.

Barn owl
A large decline occurred in the last 50 years. The main conservation actions are the creation of rough grassland habitats within lowland farmland. This can

be achieved through: (i) policy measures such as ESAs, set-aside, extensification—such areas created would have to be grazed to prevent scrub encroachment; and (ii) promoting (i) through provision of advisory material for farmers. Supporting actions need to include tighter controls on the use of second generation rodenticides and prevention of loss of nest sites — particularly old trees in eastern England and rural buildings in western Britain.

Chough

The maintenance of low intensity pastoral systems is crucial. Two particular issues need to be addressed: (i) loss of cliff-top pasture through scrub regeneration due to fencing and subsequent loss of grazing; (ii) loss of fields adjacent to cliff tops from conversion to arable or intensive grass management (e.g. re-seeding). Support for traditional pastoralism through agricultural policy measures is the most significant action, including designation of ESAs.

Quail

Unpredictable in annual numbers. Its arrival in Britain and knowledge of actual nesting areas and breeding success are virtually nil. In good years, quail frequently occurs in areas of dry grassland but also arable crops. Such measures as ESAs, extensification and set-aside may incidentally help this species.

Woodlark

About 50 per cent of the population occurs in forestry plantations and 50 per cent on heathlands. The recent increase of the species in England owes much to the re-colonisation of forestry clear-fells and re-stocks up to seven years old. The population is thus vulnerable to changes in overall forest-age structure. Requirements of woodlarks on grass and *Calluna* heathland in Britain are currently under investigation but the maintenance of heavy grazing by rabbits, sheep, cattle and horses is crucial in providing suitable feeding areas in these locations. In Breckland grass heaths, a marked decline of woodlark populations was noted with the reduction and cessation of grazing. The main measure necessary to remedy this is the creation of ESAs.

REFERENCES

Fuller, R.J. (1982). *Bird Habitats in Britain*. T. & A.D. Poyser, Calton.
Batten, L.A., Bibby, C.J., Clement, P., Elliott, G.D. and Porter, R.F. (1990). *Red Data Birds in Britain*. T. & A.D. Poyser, London.
Taylor, J.P. and Dixon, J.B. (1990). *Agriculture and the Environment: Toward Integration*. RSPB, Sandy.

APPENDIX 1

CRITERIA FOR SELECTION OF SPECIES IN "RED DATA BIRDS IN BRITAIN"

International significance of British populations
More than 20 per cent of the western European population breeds or winters in Britain.

Rare breeder
Fewer than 300 pairs breed regularly in Britain.

Declining breeder
Species whose breeding numbers have declined by more than 50 per cent since 1960.

Vulnerable breeder or non-breeder
Species often confined to rare and vulnerable habitats with more than half the population occurring at 10 or fewer sites.

IMPLICATIONS OF INTERNATIONAL LAW FOR CONSERVATION OF LOWLAND DRY GRASSLANDS IN EUROPE

Richard Buxton

Environmental lawyer; 40 Clarendon Street, Cambridge CB1 1JX, UK

ABSTRACT

A framework of potentially effective and enforceable international law exists in Europe, and particularly the European Community, which can be used to help the conservation of lowland dry grasslands. This law, however, operates merely as an element in a very complex system of checks and balances where agricultural law and policy has an enormous influence. Probably the most effective legal instrument from a conservationist point of view is a 1985 EC Directive on environmental impact assessment. A proposed EC Directive aimed at conservation of habitats, specifically including natural grasslands, is potentially powerful. The mere existence of law however is seldom enough to do any good. The challenge is to put the law into practice, something which requires constant vigilance and political pressure. Court action may be effective if opportunities arise.

INTRODUCTION

No law is yet devoted specifically[1] to the conservation of grasslands. However, as with most other environmental problems, there is already plenty of law which can be used for this purpose. The difficulty is making sure that happens.

The law that may protect grasslands and other features of the natural environment has to exist alongside law backing policy which tends to support harmful development. The working of law is seldom completely straightforward in any walk of life. Hardly ever does it promote a particular course of action with no constraints. Instead, it exists in a system of checks and balances. In this context, law designed for the purpose of protecting the environment tends to be at a disadvantage compared with the law governing other activities. It is a relative newcomer, and often operates so as to try and check well established economic interests. The people who champion the use of environmental law are at a further disadvantage as they are usually less well funded than the interests they wish to see checked.

It is, of course, agriculture which tends to have the major impacts on grasslands which conservationists wish to check. This paper is presented to complement another at this seminar (by David Baldock) which states how agricultural policy affects grasslands in the European Community. Even though it has begun to account for conservation objectives, as might be expected it is the interests of the agricultural industry which still drive agricultural policy and the law which backs it.

Hence the focus of this paper is an examination of the international law which can be used to protect grasslands against agriculture (and other damaging development). As will be discussed, agricultural law itself plays a part in this protectionist scene, and in practice that part may be more important than the separate, protective, law which has its origins in the conservation movement. In an effort to see results from the law, the paper offers practical suggestions about how to make the law work better than it does presently in achieving what it sets out to do.

APPLICABLE LAW

The context
The law discussed in this paper works on its targets in different ways. It may be designed to stop people doing things which might harm the environment. Or it may make them think hard about projects before they go ahead with them. Or it may make them take positive steps for the benefit of the environment. In either of these cases there may be sanctions against people who might wish to do otherwise, or incentives for them to organise their business so that it is best for them to act to benefit conservation.

Environmental law works at two main levels: international and domestic. International agreements (often called "conventions" or "treaties") may be truly international, affecting countries in all parts of the world. Or they may just be between states in a particular region, or simply between two states. Domestic law is law in a particular state (or just part of a state). Domestic law may exist because the people and politicians of the state have demanded it. Or it may exist so the state can fulfil its obligations under an international agreement.

European Community law, which is very important for this discussion, does not fit exactly into these categories. EC law is probably best thought of as domestic law applying to member states as if they were federalised. Sometimes EC law will apply directly, in the form of Regulations, just as in a federal system. In other cases the member states themselves must enact domestic law to comply with instructions (Directives) from the EC.

International conventions are a much more useful tool in environmental management than many people think. One can argue they are ineffective, "toothless", being so difficult to enforce. It is indeed effectively impossible for an individual or group to go to court to enforce an international convention. On the other hand, the law provides a framework for party states to work towards common environmental goals.

This is especially so where a convention has some central supervisory body, or there are regular meetings of the parties. When they meet, parties like to look as though they are complying with their legal obligations. Furthermore, in order to be seen to be active, they may set up scientific and administrative bodies to handle convention issues where none had existed before. These bodies themselves provide further momentum towards making the convention effective.

International political initiatives are a further complication in the picture. Although it is not relevant to grasslands, a good example is the North Sea Conference. The Declarations from meetings of the North Sea Ministers are not "international law" in the accepted sense, but they are political expressions which may have very similar effect. Similarly the economic effects which may result from meetings of agriculture ministers determining agricultural policy are likely to be greater and more immediate than conservationist laws.

Environmental law affecting grasslands

There are several legal tools which can be used to help conservation of lowland dry grasslands - to prevent damage, to require developers to consider the likely effects of potentially damaging projects, and to take positive conservation measures. They fall into three main groups, reflecting origins in conservation, in agriculture, and in other moves towards environmental protection.

First, and the focus of this paper, are laws which originate from conservation aims. The three most significant and presently effective ones are: the Bern Convention ("the Convention on the Conservation of European Wildlife and Natural Habitats"[2]), the EC Birds Directive ("Directive of the Council of the European Economic Community on the Conservation of Wild Birds"[3]), and the EC Environmental Assessment Directive ("Council Directive on the assessment of the effects of certain public and private projects on the environment"[4]). Another international convention, the Bonn Convention ("Convention on the Conservation of Migratory Species of Wild Animals"[5]) could be important but has yet to be put properly into effect. Finally a draft EC instrument of potentially huge importance is in the pipeline. This is the proposed Habitats Directive ("Modified proposal for a Council Directive on the conservation of natural and semi natural habitats and of wild fauna and flora"[6]).

Secondly there is the vast array of domestic and international law which in one way or another affects agriculture, ultimately affecting the way farmers make their decisions about how they should use their land. Amongst this are three types of scheme backed by EC law which are designed with conservation needs partly in mind. These are environmentally sensitive areas, extensification, and set-aside. This paper summarises the principles of these schemes and how the law backs them up; a further recent development is also mentioned[7].

Thirdly there are several legal instruments designed to protect different aspects of the environment but which may have an effect on grassland health and development. At an international level these include legislation designed to curb the use of pesticides[8], to limit nitrates in drinking water[9]; and, one might argue, measures designed to help with atmospheric pollution[10]. It is beyond the scope of this paper to discuss these further, and they should not divert attention from the more important laws on which to focus and see put into practice. However, their existence, if nothing else, does illustrate the complexity of the problems and potential remedies affecting lowland dry grasslands.

A description of what these laws are, or will be, supposed to do follows. The Bern Convention applies to many European (and two African) countries[11], while the Bonn Convention has signatories worldwide[12]. EC law of course only applies to the states of the European Community. Nevertheless, Community states cover a huge area and put fully into practice Community law could have much influence on the kinds of problem this seminar addresses. So, they are worth considering properly. They are also of practical importance as Community law is more easily enforced than other international agreements, because individuals and groups can use it in domestic courts or through the European Commission.

The Bern Convention — habitat and species conservation

The Bern Convention, now eleven years old, aims at conservation of both habitat and specific flora and fauna. The basic obligation is set out in Article 2:

The Contracting Parties shall take requisite measures to maintain the population of wild flora and fauna at, or adapt it to, a level which corresponds in particular to ecological, scientific and cultural requirements, while taking account of economic and recreational requirements and the needs of sub-species, varieties or forms at risk locally.

The Convention is a very positive instrument, containing clear and unequivocal obligations on party states actually to take certain steps towards this goal. In common with others, it does contain some horatory language ("to encourage", "to promote", "to consider" etc. found in so many other conventions—for example, the Bonn Convention, considered below) but it is unusually definite in what it requires of party states.

The Convention emphasises the protection of endangered and vulnerable species, especially migratory ones. It prohibits deliberate damage to or destruction of breeding or resting places. Party states must take appropriate steps to protect animals and plants of the species listed, and maintain populations. They must promote national policies in keeping with this, including planning and development policies and measures against pollution which have regard for wildlife conservation, as well as education and information dissemination. The Convention provides a framework for uniting European countries' approach to wildlife conservation under common legal standards.

Although species protection is the underlying aim, the Convention recognises that habitat conservation is a vital component of this, and so contains certain provisions which are specific to habitat. Article 4 states:

1. Each Contracting Party shall take appropriate and necessary legislative and administrative measures to ensure the conservation of the habitats of the wild flora and fauna species, especially those specified in the Appendices I and II, and the conservation of endangered natural habitats.

2. The contracting parties in their planning and development policies shall have regard to the conservation requirements of the areas protected under the preceding paragraph, so as to avoid or minimise as far as possible any deterioration of such areas.

3. The Contracting Parties undertake to give special attention to the protection of areas that are of importance for the migratory species specified in Appendices II and III and which are appropriately situated in relation to migration routes, as wintering staging, feeding, breeding, or moulting areas.

4. The Contracting Parties undertake to co-ordinate as appropriate their efforts for the protection of the natural habitats referred to in this Article when these are situated in frontier areas.

Article 6, which is mainly species related, states:

Each Contracting Party shall take appropriate and necessary legislative and administrative measures to ensure the special protection of the wild fauna species specified in Appendix II. The following will in particular be prohibited for these species: .... b) the deliberate damage to or destruction of breeding or resting sites....

Article 9 of the Convention allows parties to make exception from these rules "provided that there is no other satisfactory solution and that the exception will not be detrimental to the survival of the population concerned". The Article is drafted so as to limit it being used as an excuse for states not taking their obligations seriously: it specifies when the exceptional activities can take place (for example, "in the interests of public health and safety, air safety or other overriding public interest") and requires parties to report to the Convention's Standing Committee with details of the exceptions made.

Part of the Convention's strength is in its administrative system, which involves this Standing Committee to which parties must report, and which is in turn responsible to the Council of Europe, which houses a permanent Secretariat. This structure works to keep parties "up to the mark". Meetings of the Standing Committee, which government and non-government representatives attend, can be good opportunities to put pressure on governments to have them take action on problems. Procedures also exist for direct appraisal of a site by representatives of the Standing Committee to which party states have to answer.

Dry grasslands have been specifically identified by the Standing Committee of the Bern Convention as particularly vulnerable habitat types. In 1984 the Committee recommended parties to draw up inventories of these habitats identifying those which "receive protection which ensures their conservation"[13]. According to the Bern Convention Secretariat, this recommendation had a "very poor response" and there has been no further action on dry grasslands within the Convention framework[14]. (There was a good response in connection with heathlands, which was part of the same recommendation. Perhaps the poor response on grasslands was partly due to difficulties in precisely defining this habitat type.[15])

There is special significance in the recent joining of two states to the Convention. Senegal set a precedent for African countries, and recognises the importance which conservation in Africa has to play in protecting European wildlife. Senegal has now been joined by Burkina Faso. Hungary's joining is important for grasslands, being the gateway to Europe from the Russian steppes. It was the first of what may be several other eastern European and north African states: Bulgaria has very recently (February 1991) joined the Convention; Poland and the USSR are likely to do so during 1991; and Czechoslovakia, Yugoslavia, Morocco, and Tunisia have all expressed an interest in the Convention and have been invited to join[16].

EC Birds Directive — Bird habitat and species protection

The EC Birds Directive came into effect in 1981. It imposes strict, legal requirements on member states to maintain populations of naturally occurring wild birds at levels corresponding to ecological requirements, to preserve sufficient diversity and area of habitat for their conservation, to limit hunting, to regulate trade in birds, and to prohibit certain methods of killing and capture.

The obligation to maintain populations apparently comes before economic and other considerations. Article 2 states:

Member States shall take the requisite measures to maintain the population of the species referred to in Article 1 at a level which corresponds in particular to ecological, scientific and cultural requirements, while taking account of economic and recreational requirements, or to adapt the population of these species to that level.

This is followed in Article 3 by habitat conservation obligations:

In the light of the requirements referred to in Article 2, Member States shall take the requisite measures to preserve, maintain or re-establish a sufficient diversity and area of habitats for all the species of birds referred to in Article 1.

The preservation, maintenance and re-establishment of biotopes and habitats shall include primarily the following measures:
  (a) creation of protected areas;
  (b) upkeep and management in accordance with the ecological needs of habitats inside and outside the protected zones;
  (c) re-establishment of destroyed biotopes;
  (d) creation of biotopes.

Article 4 of the Directive builds on these requirements in relation to birds listed in Annex I to the Directive, which require special conservation measures. Member States are obliged to classify suitable areas for special protection for the conservation of these species, and take steps to avoid pollution or deterioration of habitat.

The European Commission is ultimately responsible for compliance with the Directive. Member states must report to it. The Commission can (and has) taken proceedings in the European Court against states which fail to enforce the Directive. Such failure might mean not properly putting the Directive into effect in domestic law and policy, or not acting against citizens and organisations who do things which breach the Directive. However, as an interim judgment in the case referred to in the next paragraph showed, the European Commission does have difficulty in acting decisively in a particular case in order to prevent damage to habitat.

That case, however, ultimately confirmed the strength of states' obligations under the Directive. In particular the European Court said that Article 4(4) gives very little scope for allowing deterioration of designated special protection areas. The court confirmed that derogation can be justified only on exceptional grounds, which must be an interest superior to the ecological objective of the Directive, and that economic and recreational requirements are not sufficient[17].

The EC Environmental Assessment Directive — Screening of projects for potential environmental effects

The EA Directive requires a formal statement ("environmental statement" or "ES") about the potential impact of projects which may significantly affect the environment before consent for the development project is given. One must assess the likely environmental impact, and identify measures to avoid or reduce adverse effects. The scope of the Directive is very wide. According to Article 1:

This Directive shall apply to the assessment of the environmental effects of those public and private projects which are likely to have significant effects on the environment.

Article 2 defines "project" to mean:

The execution of construction works or of other installations or schemes, or other interventions in the natural surroundings and landscape including those involving the extraction of mineral resources.

Some types of projects must be assessed for their potential environmental effects. These are the so-called Annex I projects which tend by their nature to be large scale, and where it is "obvious" that significant environmental effects will occur - for example, construction of power stations, radioactive waste disposal sites, and larger trading ports. Then there is a much larger grouping of project types, listed in Annex II of the Directive, where there must be assessment where EC Member States "consider that their characteristics so require"[18].

Defence projects and projects approved by specific legislation are exempt from the requirements of the Directive. The use of "specific legislation" as a means, intentionally or for other reasons, of avoiding EA as a legal requirement is a serious problem. The private bill procedure in the U.K. is a well known example[19]. There are also, for example, apparently similar problems in Portugal[20].

Many of the Annex II project types could conceivably affect grasslands, but two are very pertinent, coming under the heading "agriculture":

(b) Projects for the use of uncultivated land or semi-natural areas for intensive agricultural purposes.

(d) Initial afforestation where this may lead to adverse ecological changes and land reclamation for the purposes of conversion to another type of land use.

The Directive is supposed to work by EC Member States integrating it into their existing development consent procedures. So, for example, in the U.K., the Directive is implemented by several sets of Regulations (secondary legislation made by Statutory Instrument) which deal with the different development consent procedures relating to regulation of town and country planning, harbours, afforestation, and so forth. A particular problem so far as grasslands are concerned is that in most cases in the U.K. agriculture is exempt from development control[21] so that grassland which might be affected by it effectively remains unprotected by EA requirements.

To counter this there is a potentially helpful provision in Article 2.2 of the Directive, which states:

The environmental impact assessment shall be integrated into the existing procedures for consent to projects in the Member States, or, failing this, into other procedures to be established to comply with the aims of the Directive [emphasis added].

How might one see this provision put into practice? Like other Directives, the EA Directive is supposed to be put into practice by appropriate domestic legislation in individual member states. However, it is becoming increasingly usual for domestic courts to find that Directives themselves may have "direct effect" - in other words, apply in a particular country as domestic law without any need for domestic legislation being passed. To enable this to happen, the wording of the Directive must be sufficiently clear and unambiguous.

The EA Directive is particularly interesting in this way because of the clear, unambiguous, and mandatory language that it contains. One can, as it were, focus it nicely on an individual development project, and ask the relatively simple question, "have the clear words of the Directive been complied with or not?"[22]. (By contrast, other instruments like the Birds Directive aim at requiring states to take much more general action, for example to conserve habitat, and it is much more difficult to complain about their failure to do so in the context of a specific case.)

Conservation of grasslands might be served a good cause by someone challenging a government to introduce development consent procedures for agricultural work where that is likely to have significant environmental effects, on the basis of the requirements of Article 2. There are also moves in the European Commission to reinforce implementation of the requirements of the Directive on agricultural projects, and, apparently, add to the list[23].

EA may not stop a project, but it is at least a way of forging good compromise. It can help put developments where they should be, and keep them out of where they should not be. And the process can make sure that when the form of the project is agreed, the work harms the environment as little as possible, or even benefits it.

The Bonn Convention — protection of migratory species

The fundamental aim of the Bonn Convention is to protect migratory species. The Convention works in two ways. First, it imposes strict conservation obligations on states which form part of the range of specially threatened species (listed in Appendix I to the Convention). Secondly, it obliges states which form part of the migratory range of species under lesser threat (listed in Appendix II) to conclude agreements which will benefit the conservation and management of these species.

The Convention obliges states directly (Appendix I cases) or as part of agreements with other states (Appendix II cases) to conserve and, where feasible and appropriate, restore important habitats of the species in question. The language of the Convention is very similar in both cases, but unlike some instruments - for example, the Bern Convention, or the EC Directives mentioned - is couched in softer, less mandatory, language.

Potentially the Bonn Convention could be of indirect value to grasslands, by virtue of the habitat provisions mentioned. To date, there have been negotiations between states in respect of four species types - small cetaceans in the North Sea and Baltic Sea, European bats, white storks (*Ciconia ciconia*), and western palearctic waterfowl) but none of these have yet resulted in concluded agreements. Presently it is unclear how the agreement system is in fact supposed to operate. Some parties believe agreements should be formal agreements between governments, with treaty status, and therefore quite complicated and time-consuming to ratify and put into effect. Others take the view that agreements with simple ministerial signatures will be adequate - provided that can enable proper funding of the conservation measures agreed. It is possible that these problems will be ironed out over the coming months with conclusion of the agreement in relation to European Bats, and with that precedent other agreements may fall into place more easily[24]. Another problem is that meetings of the parties take place only every three years, less frequently than with e.g. the Bern Convention.

So far as lowland dry grasslands are concerned, the species presently considered for conservation and management agreements would have little impact on this type of habitat, except possibly in the case of white storks. There have however recently been very preliminary discussions between Italian and Hungarian officials in connection with conservation of the great bustard (*Otis tarda*) within the framework of the Convention[25]. There is also a meeting of the parties in October 1991 and it may be that this could be used to put forward a case for an agreement in respect of other migratory species where lowland dry grasslands are central to habitat requirements.

Proposed Habitats Directive — protection of general wildlife habitat

The proposal for a Habitats Directive arose from the EC's Fourth Environmental Action Plan, which recognises the need for protecting not just birds, or endangered species, but the natural and semi-natural habitat of animals and plants generally. Only in this way, it is argued, can one hope to fulfil the main objectives of the World Conservation Strategy - to maintain essential ecological processes and life support systems, to preserve genetic diversity, and to use species and ecosystems in a sustainable way.

Article 2 states:

1. Member States shall take the requisite measures to maintain the abundance and diversity of wild fauna and flora at a level which corresponds in particular to ecological, scientific and cultural requirements and the needs of sub species, varieties, forms and populations at risk locally, while taking account of economic and recreational requirements.

2. Member states shall take appropriate steps to monitor the conservation status of species and habitats mentioned in Article I in all the regions of their territories where they occur, taking particular account of
(a) the need to monitor the status of threatened species and habitats, and
(b) the need to monitor the effectiveness of measures undertaken pursuant to paragraph 1 of the present Article"

In relation to habitats, articles 4 and 5 provide:

Member States shall take the requisite measures to conserve the natural and semi-natural habitats of wild fauna and flora, with particular attention to the habitats of threatened species, especially threatened species endemic to the European territory of the Member States and threatened natural and semi natural habitats, in accordance with the provisions of this Directive.

...the types of habitat specified in accordance with Annex IV shall be the subject of special conservation measures in order to ensure the maintenance or re-establishment of the species concerned at a satisfactory conservation status in their areas of natural distribution as well as conservation of the habitats concerned in all the regions where they occur.

Annex IV lists the "threatened" habitats and these include various types of natural and semi-natural grassland (including, but not specifically, the grasslands subject of this seminar)[26]:

Natural grasslands:
34.1	Karstic "barren" grasslands (34.11) or xeric sands (34.12)
34.2	Calaminarian grasslands (rare and endemic, in central Europe)
36	Alpine and boreal grasslands
36.314	Siliceous *Festuca eskia* grasslands of the Pyrenees (endemic)
36.32	Siliceous alpine and boreal grasslands of the Scottish Highlands (endemic)
36.36	Siliceous *Festucetea indigestae* Iberian grasslands
36.4	Alpine calcareous grasslands: all types from 36.41 to 36.45
36.5	Macaronesian sub-alpine grasslands

Abandoned former grazing land (and facies where bushes grow):
35.1	On siliceous substrates
35.11	Hautes Chaumes (Vosges, Black Forest, Jura, etc.)
35.12	Sub-mountainous (*Nardus* grasslands)
34.3	On calcareous substrates (*Festuco-Brometea*) (sites of remarkable orchids) (34.33, 34.34, 34.35, and 34.36)
34.5	Pseudo-steppe with grasses and annuals

Sclerophyllous grazed forests (*dehesas* in Spain):
32.11	*Quercus suber* and/or *Q.Ilex*

Semi-natural tall herb grasslands:
37	All types especially *Molinion* (37.3 and 37.4)

Humid grasslands with *Cnidion venosus*

Annex VII lists landscape features of outstanding local importance to wildlife. Member states are obliged by Article 8(2) to ensure protection of distinct landscape features, including those listed in Annex VII, which are of outstanding importance to wildlife. The following listed features are relevant in the context of grasslands conservation:

Treelines
    Grassy terrace slopes
    Scrub patches
    Small woodlands
Ponds, temporary ponds, and waterholes
Humid depressions
Dry uncultivated hilltops
Arable field margins
Stony areas
Herbaceous layer of orchards and plantations
Edges of waterways and water-bodies
Salt pans
Valley corridors

Under the proposed Directive, Member states would be obliged to classify special protection areas and then take "appropriate steps" to avoid pollution or deterioration of habitats or significant disturbances to fauna and flora in those areas. States should create integrated management plans to suit the ecological needs of the special protected areas. In order to ensure "a satisfactory conservation status of a species" states "shall envisage the re-establishment of destroyed or degraded biotopes or the creation of new ones"[27]. States are also supposed to do their best to avoid pollution and protect habitats outside the classified areas. There are detailed provisions relating to protection of particular species of animals and plants; for providing information about implementation of the Directive; for encouraging research; and for monitoring populations and communities.

As the Directive is presently drafted, each state must classify ten of its best areas for special protection within two years of the Directive coming into force, failing which the European Commission will do this for them. Coverage should be complete within eight years. By then, there must be territories in each member state large and numerous enough to ensure (i) the maintenance of specified species at a satisfactory level in all regions where they occur and (ii) the protection of threatened habitats and associated flora and fauna in all regions where they occur. This provision is one of the most contentious in the negotiation of the Directive: member states are sensitive about its implications for sovereignty, and also the proposed designation requirements are seen as inappropriately inflexible. It is likely that the provision will be replaced with something more flexible, which will nevertheless achieve satisfactory conservation status for species and habitats in regions where they occur[28].

The Directive envisages close cooperation with and supervision from the European Commission in putting the Directive into practice. The ultimate aim is to establish a coherent network of areas across Europe ("Natura 2000"). These areas would include special protection areas classified under the Birds Directive.

The proposed Habitats Directive would also link with the Environmental Assessment Directive. It would need to be amended to make it clear that development programmes and specific projects which are within, or likely to affect, the conservation potential of a special protection area must be subject to environmental assessment. People involved with the protection and management of the special protection areas must be closely associated with the assessment procedures and their "final opinion" must be "made public".

The proposed Habitats Directive is presently in draft form. It has not yet become law. Provided states can be satisfied that funds (including those directed at compensation to farmers etc. for lost/foregone profits) will be put in place to make it work, it is likely to become law by the end of 1991[29]. It would come into force two years after that.

Agricultural legislation

The principal EC measures which give legal effect to agricultural practices which may benefit grasslands are regulations providing for designation of environmentally sensitive areas, extensification, and set-aside. One cannot repeat too often that these operate merely as a part of the wider agricultural scene. Difficulties for conservationists are expressed nicely in the preamble to the Regulation: it first encourages cut backs in agricultural production, and then, in the next paragraph, encourages afforestation of agricultural land - something which is not always to the benefit of conservation.

Whereas farmers in areas that are sensitive from the point of view of protection of the environment or preservation of the landscape are in a position to perform a valuable service to society as a whole and whereas the introduction of specific measures may encourage farmers to introduce or retain agricultural production practices that are compatible with the increased need to protect or preserve the countryside and at the same time help, by means of reorientation of their holdings towards realisation of the agricultural policy objective of restoring balance on the markets in certain products;

Whereas the measures to encourage the afforestation of agricultural land must be amplified[30]

The dichotomy is expressed more fundamentally in the Treaty of Rome itself, which has as one of its founding tenets Article 39 stating the objectives of the Common Agricultural Policy:

(a) To increase agricultural productivity by promoting technical progress and by ensuring the rational development of agricultural production and the optimum utilisation of the factors of production, in particular labour ...

This is now tempered by Article 130R, which provides:

Environmental protection requirements shall be a component of the community's other policies.

The intent of the preamble to Regulation 1760/87 quoted above in relation to environmentally sensitive areas is put into effect by Title V, Article 19 which is headed

Aid in areas sensitive as regards protection of the environment and of natural resources and as regards preservation of the landscape and the countryside

and goes on to provide as follows:

In order to contribute towards the introduction or the maintenance of farming practices compatible with the requirements of the protection of the environment and of natural resources or with the requirements of the maintenance of the landscape and the countryside, and thus to contribute to the adaptation and the guidance of agricultural production according to market needs and having regard to agricultural income losses resulting from this, Member States may introduce a specific aid scheme for areas which are particularly sensitive from these points of view.

The aid scheme ... shall consist of an annual premium per hectare granted to farmers ... who undertake, under a specific programme for the area concerned, to introduce or maintain, for at least five years, farming practices compatible with the requirements of the protection of the environment and of natural resources or with the requirements of the maintenance of the landscape and of the countryside.

[Article 19 goes on to provide for member states to determine the relevant areas, appropriate production practices, levels of farming intensity and/or livestock density, and amount of premium, subject to a maximum.]

Environmentally sensitive area schemes are, as the text quoted indicates, designed to modify farming practices to lower intensity of production or to discourage intensification. These have been followed by the closely related concepts of extensification and set-aside. In order to benefit from the environmentally sensitive areas scheme, one has to be in an area designated as an ESA.

Extensification on the other hand involves sectoral payments (for example, to the beef, veal, or milk sectors), in other words relating to the *type of farming* in question rather than specific to an *area*. The effect is similar, in encouraging the farmer to convert to less intensive farming systems. The legal authority for extensification lies in Regulation 1094/88[31], which provides:

Member States shall introduce an aid scheme to promote extensification for surplus products. ...... Extensification shall be defined as a reduction of at least 20%, for a period of at least five years, in the output of the product concerned without any increase in other surplus production capacity

Set-aside, which is authorised by the same Regulation 1094/88, is to enable farmers simply to retire a proportion of their land from production, by paying compensation for the production foregone. It applies only to land which has already been used for production of crops, and the farmer is obliged to keep the land in good agricultural condition, "with a view to protecting the environment and natural resources"[32]. Set-aside is thus hardly useful for grasslands which have never been cultivated[33], but could be useful in situations where arable land might be returned to grassland.

Recently a proposal has been published for a Council Regulation on the introduction and maintenance of agricultural methods compatible with the protection of the environment and the maintenance of the countryside. It would promote traditional and environmentally friendly agricultural practices, decreased use of fertilisers, reduction of yields, and reduction of soil erosion[34].

Putting law into practice

Even if at times the law competes, conflicts, and overlaps, it is clear that there is a great deal of it which can affect the conservation of European lowland dry grasslands. The Bern Convention, and the EC Birds and proposed Habitats Directive, all aim, with differing emphasis, on protection of wildlife habitat and individual species. If and when the Habitats Directive becomes law in the European Community and is implemented in the domestic law of the member states, grasslands will have specific, tough, and positive protection. EC agricultural policy, backed by legal regulation, offers encouragement to conservationist farming practices. A mass of other legislation, which here has only been mentioned in passing, also has an impact. Finally the Environmental Assessment Directive offers a powerful check on decision making until the adverse consequences of developments are properly considered.

So, the outlook for grasslands should look good. However, despite the plethora of legal tools available to promote grasslands conservation, it would be wrong to think that the job is done, and that the law will somehow work by magic. The fact is that law is only so good as the wills of individuals, states, and, in the case of the EC, the European Commission, to see that the law is used as intended.

So far as the law which has been referred to here as originating from the conservation side is concerned, that means that individuals and conservation organisations must keep constant pressure on their governments to take the steps the law requires them to take and constant vigilance against failure to do so. Governments have many competing priorities. So pressure from individuals and organisations below, and from the European Commission above, is the best, if not the only, way to see EC and other international law put properly into practice. The European Commission can act on its own initiative but is in practice much more likely to do so in response to complaints and pressure from individuals and organisations in member states. So whatever the law may say, the initiative in practice lies with individuals - particularly the people at this meeting.

The same is obviously less true for agricultural schemes, as this works by influencing the business decisions of individual farmers. However, there is still the opportunity to lobby governments to put such schemes into place where they do not exist. On a much more local scale one may be able to point out to individual farmers, who may not be aware of them, the opportunities that these schemes offer them. No doubt farmers will be much more receptive to news that money is available to them under a particular scheme, than news that they have to comply with some conservationist law which they may find hard to understand.

Finally there is the difficult question whether individuals and organisations can themselves take action in the courts to enforce the law directly. How difficult and expensive that is varies between different countries. The rules relating to title to sue, or *locus standi*, are complex[35]. There are some signs that the EC will eventually harmonise these rules to allow more freedom for people to take action than they presently have in some member states[36]. It may be very difficult through the legal process to force states to take positive steps to comply with international law (for example, create special protected areas), although the doctrine of direct effect of EC Directives is becoming more widespread.

Apart from alleging that a particular Directive has direct effect, there may often be legal redress available in cases where a particular action has been carried out or a decision made which is contrary to the law. For example, in the U.K., where a public authority gives consent for a project which should, but does not, require environmental assessment, an individual or organisation may be able to object in the courts by a process known as judicial review. The decision of the court will relate to the case in question, but may force the issue more widely at government level.

CONCLUSION

The aim of this paper is to show that Europe, and particularly the European Community, is well endowed with law which can be used to protect grasslands. The task of people interested in conservation is to press for that law to be put properly into practice. The proposed EC Directive on habitats would add particularly usefully to the range of law available and therefore it will be good for lowland dry grasslands if and when this becomes law in the Community.

NOTES

[1] See, however, the provisions of the proposed EC Habitats Directive, discussed later, which specifically includes grasslands.
[2] Signed at Bern, Switzerland, 19th September 1979
[3] Council Directive 79/409/EEC (2.4.89)
[4] Council Directive 85/337/EEC (5.7.85)
[5] Signed in Bonn, June 1979
[6] OJ COM (91) 27 (final) 8.2.91
[7] Proposal for a Council Regulation on the introduction and maintenance of agricultural production methods compatible with the requirements of the protection of the environment and the maintenance of the countryside: Official Journal, C.267/11 1990 (November 1990)
[8] E.g. Directive 76/896/EEC and amendments (fixing of maximum levels for pesticide levels on fruit and vegetables); 86/363/EEC (fixing of maximum levels for pesticide levels for pesticide residues in and on foodstuffs of animal origin)
[9] 80/778/EEC
[10] E.g. Regulation 3528/86 and successors
[11] Austria, Belgium, Bulgaria, Burkina Faso, Cyprus, Denmark, the EC, Finland, France, Germany, Greece, Hungary, Ireland, Italy, Liechtenstein, Luxembourg, the Netherlands, Norway, Portugal, Senegal, Spain, Sweden, Switzerland, Turkey, and the U.K.
[12] Belgium, Benin, Burkina Faso, Cameroon, Chile, Denmark, Egypt, Finland, Germany, Ghana, Hungary, India, Ireland, Israel, Italy, Luxembourg, Mali, Netherlands, Niger, Nigeria, Norway, Pakistan, Panama, Portugal, Saudi Arabia, Senegal, Somalia, Spain, Sri Lanka, Sweden, Tunisia, United Kingdom, Uruguay, Zaire, European Economic Community
[13] Recommendation No. 3 (1984)
[14] E. Galiano, Bern Convention Secretariat, pers. comm.
[15] *Ibid.*
[16] *Ibid.*; L. Rose, RSPB pers. comm.; S. Lyster, WWF-UK pers. comm.
[17] European Commission v Federal Republic of Germany, re: the Leybucht Dykes; Eurpean Court of Justice, unofficial judgment 28.2.91. Earlier judgment at [1990] 3 CMLR 651
[18] Directive EEC/85/337, Article 4.2
[19] EA may however shortly be required in these cases: see Department of the Environment consultation paper, "Environmental Assessment and Private Bill Procedures", 26.2.91
[20] Wildlife and Development in the European Community, Royal Society for the Protection of Birds, December 1990, p. 9
[21] Town and Country Planning Act 1990, s. 55
[22] To what extent the EA Directive may have direct effect is at the time of writing a confused issue. It seems accepted that projects for which EA is mandatory under Annex I of the Directive would have direct effect, but according to a recent Scottish decision, Re: petition of Kincardine and Deeside District Council for a review of a decision of the Forestry Commissioners: [judgment 8.3.91, unreported], Annex II projects cannot have such effect. The position may be clarified on appeal or by a reference to the European Court.
[23] New Scientist, 23.3.91
[24] D. Hykle, Bonn Convention Secretariat, pers. comm.; S. Lyster, WWF-UK, pers. comm.
[25] J. Johnson, Bonn Convention Secretariat, pers. comm.
[26] This is part of Annex IV of the draft Habitats Directive, titled "Natural and semi natural grassland formations". It is one of nine habitat types designated for protection in the EC. Numbers refer to Corine classification.
[27] Draft Directive, Article 5
[28] I. Hepburn, RSPB, pers.comm

[29] I. Hepburn, RSPB, pers. comm.
[30] Regulation EEC 1760/87 (amending Regulations 797/85, 270/79, 1360/78, and 355/77 as regards agricultural structures, the adjustment of the countryside to the new market situation and the preservation of the countryside)
[31] Council Regulation of 25th April 1988 amending Regulations (EEC) No. 797/85 and (EEC) No. 1760/87 as regards the set aside of arable land and the extensification and conversion of production
[32] *Ibid.* Article 1a(3)
[33] Land growing crops not covered by a common market organisation are excluded from the scheme: *Ibid.* Article 1a(2)
[34] OJ C.267/11 1990
[35] In the UK, one is supposed to have "sufficient interest" in a matter. That means more than being a mere member of the public who happens to be interested in an issue. On the other hand, it is not necessary to have a direct property or personal interest. It is not inaccurate to suggest that the greater the merits of a case, the less likely the court is to reject one on grounds of standing. It is obviously an issue to be considered in every case, but poor standing may not be a fatal block to pursuing the case. The issue may not even be raised by the other party.
[36] See Proposal for a Council Directive on civil liability for damage caused by waste, COM(89)282 final 4.10.89 OJ C/251/3

IMPLICATIONS OF EC FARMING AND COUNTRYSIDE POLICIES FOR CONSERVATION OF LOWLAND DRY GRASSLANDS

David Baldock

Institute for European Environmental Policy, 3 Endsleigh Street, London WC1H 0DD, UK

INTRODUCTION

The great majority of grassland in Europe is under some form of agricultural management and this is true also of lowland dry grasslands. The nature of this agricultural management is clearly of central importance in determining the composition and extent of such grassland. Accordingly, this paper will be devoted mainly to a discussion of agricultural and related policies pursued by the European Community and some of their potential implications for grassland, both now and in the longer term.

If agricultural statistics were to be recast in a biological mould, land use in different EC countries would be described in rather different terms than are used at present. For example, grassland might be broken down into categories such as acidic, basic and neutral and official statistics might distinguish between wet and dry, lowland and upland grassland. For the moment, however, most land use statistics reflect the commonplace agricultural forms of classification in which the main distinctions are between permanent and temporary grassland and in Britain between the broad category of permanent grassland, predominantly in the lowlands, and "rough grazing", the majority of which is in the uplands.

Agricultural policy traditionally has paid little regard to the different categories of grassland utilised by biologists and conservationists. This is one reason why it is difficult to assess the impact of Community agricultural policies on grassland with any precision. At a more general level, it is possible to trace significant changes in broad land use patterns in different countries. These can be supplemented by the results of more local studies and detailed investigations of particular habitat types. However, there is not sufficient data to allow for a detailed analysis of precisely how the various Community agricultural policies impinge on specific categories of grassland. Unavoidably, much of this paper will be concerned with grasslands as a whole, with selective references to specific types of grassland and lowland dry grassland in particular.

OVERALL TRENDS IN GRASSLAND MANAGEMENT

At present, there is no satisfactory system for monitoring changes in different habitat types in most European Community countries; still less is there a system for monitoring such developments at a Community level (Baldock 1989). Consequently, while it is clear that the total area of many semi-natural grassland habitats has declined sharply in Europe, much of the information available is very fragmentary and not comparable between countries. Dry grasslands in Europe were the subject of a report by the European Committee for the Conservation of Nature and Natural Resources published in 1981 which indicated that most countries had experienced substantial declines in the remaining area since the Second World War (Wolkinger and Plank 1981). In many cases agriculture was identified as the principal cause of decline. For example, in Denmark use of fertilisers and agricultural improvement was an important factor, while in France there was specific reference to land consolidation. Damage from agriculture is by no means confined to the European Community; over-grazing has been a major cause of damage in Sweden for example. Many of the remaining areas of dry grassland occur in protected areas, including nature reserves and national parks.

Statistics are available for the overall area of pasture, meadow and other agriculturally managed grassland in the EC countries. While the definitions of grassland tend to vary considerably between countries and are not necessarily comparable especially between Mediterranean and northern European countries, it is possible to obtain an overall view of the direction of change. Table 1 shows the total area of permanent grassland in the twelve EC

Table 1: Permanent grassland in EC countries*, 1986

Country	Area of permanent meadow/pasture (000 ha)	%of national agricultural area	Rate of decline 1980-88
Belgium	612	44.5	-0.9
Denmark	217	7.5	-1.8
Fed. Rep. Germany	4,449	37.3	-0.8
Greece	1,789	31.2[2]	0.0
Spain	6,650	24.5	-0.1
France	11,469	37.9[1]	-1.3
Ireland	4,666	81.6[1]	0.3
Italy	4,858	28.4[1]	-0.7
Luxembourg	69	54.7[1]	-0.3
Netherlands	1,081	53.5	-0.9
Portugal	761	16.8[1]	0.0
United Kingdom	11,586	62.5[1]	-0.3
Europe 12	**48,487**	**38.0[2]**	**-0.5**

* As defined in agricultural statistics
1: 1987 or 1987/1979
2: 1986 or 1986/1979
Source: Eurostat

countries, drawn from EC statistics (Commission of the European Communities 1990). Much the largest areas of agricultural grassland are to be found in the UK and France and these two countries, together with Spain, account for more than half the total area in the Community.

In most countries the overall area of grassland is in decline, with a crude Community average of 0.5 per cent per annum in the period 1980-88. During the same period, the overall area of arable land in the Community has been stable, while the total area of agricultural land as a whole has been falling by an average of 0.2 per cent per annum. This suggests that land has been moving out of permanent grass into arable production during this period. Such a trend can be explained partly by relative prices for arable crops and livestock products. For example, there was a substantial loss of grassland to arable after the UK entered the European Community in 1973 and cereal prices rose to align with the CAP price regime.

However, trends within particular countries vary considerably. In Luxembourg, for example, the introduction of the CAP resulted in lower prices for cereals relative to the previous national regime and this lead to the total area of grassland growing significantly at the expense of cereal production (Frisch 1988).

In France, the total area of grassland has declined fairly rapidly during the 1980s with considerable areas of pasture being converted to arable land on the one hand, while poorer grassland has been abandoned altogether in areas such as the Auvergne and the Pyrenees, leading to new areas of *friche* (recently abandoned land) and woodland resulting from natural succession.

In the UK, figures for habitat loss, including limestone grassland etc., are well known and will not be repeated here. Despite the considerable efforts to introduce conservation objectives into agricultural policy and the growing extent of protected areas, grassland continues to be subject to threat from agricultural intensification. According to the Nature Conservancy Council, there are three principal problems. One is the "improvement" of grassland which occurs by means of drainage, reseeding, increased fertiliser use, etc. and reflects greater pressures on farmers to increase yields. Second is the drainage of wet grasslands and third is neglect and under-grazing arising partly from the decline of livestock farming in areas now devoted almost exclusively to arable production. This has resulted in small areas of grassland which are not suitable for cultivation, being partly or wholly abandoned and becoming vulnerable to scrub invasion (Nature Conservancy Council 1991). Abandonment of grazing occurs even in prosperous intensively farmed regions, such as Kent in southeast England (Green 1990).

Similar pressures exist in other northern European countries, with some variation depending on local conditions and land use policies. For example, land consolidation projects can result in extensive drainage and land improvement works but continues in many northern European countries including France, the Netherlands and Germany. In Spain and Portugal, abandonment of traditional grazing patterns is a major threat to some areas of semi-natural grassland and agro-pastoral systems such as the *dehesas* in Spain. As traditional pastoral systems are becoming increasingly incapable of generating sufficient income to ensure their viability, they become more vulnerable to alternative land uses, such as commercial afforestation, which is often subsidised either by national agencies or the European Community or both.

In summary, the main pressures on semi-natural grassland habitats from agriculture arise either from intensification or from land abandonment and transfer into other uses. Some of the policies underlying these processes are outlined briefly below.

EC POLICIES ASSOCIATED WITH INTENSIFICATION AND LAND ABANDONMENT

Changes in agricultural practice cannot be attributed wholly to public policy, whether it be regional, national or Community in origin. Other important considerations include changes in agricultural technology, changes in the labour market, patterns of taxation and credit affecting the farming community, etc. However, agricultural policy plays an important role and is the sole concern of this paper.

The central feature of agricultural policy in the EC is the market support mechanism of the Common Agricultural Policy (CAP). In essence, this seeks to achieve a minimum price for all the most important European agricultural products, with the exception of potatoes, by means of a variable tariff on imports, a system of removing certain surplus products from the market and an associated system for subsidising the export of commodities. For most commodities, EC prices are substantially above those ruling the world market, although the latter is artificially depressed by the widespread practice of subsidising exports.

The combination of both high and stable price levels and the open-ended system of support available from the Community agricultural budget, FEOGA, has encouraged a steady increase in agricultural output and the development and adoption of a wide range of new technologies. Most economists agree that this system of price support has been one of the key factors underlying agricultural intensification and growing specialisation in European agriculture (Bowers and Cheshire 1983, von Meyer 1988, OECD 1989). Furthermore, the relative prices for different commodities under the CAP have had an important influence on land use in many EC Member States. The introduction of CAP price levels for cereals and oilseeds has led to the ploughing of grassland in areas where arable production previously had not been worthwhile.

It should be noted that the EC is by no means alone in setting high prices for agricultural commodities. The European Free Trade Association (EFTA) countries have also provided generous systems of price support for their farmers - generally at a higher level than those offered in the EC. Sweden, for example, has generated considerable surpluses of cereals and has had to subsidise their export, partly by introducing taxes on the use of fertilisers.

For milk and sugar production in the EC, the price regimes are supplemented by quota systems. In the case of milk, which is of particular relevance to grassland management, each country has a fixed production quota, which is distributed among individual farmers according to national rules which differ considerably. The impact on land use of milk quotas, which were introduced in 1984, has not been studied in depth, but is likely to vary significantly between farms. In some cases, farmers have abandoned milk production altogether, others have diversified into sheep or beef cattle production, while others have concentrated on making maximum use of grass and fodder crops grown on their own holdings in order to minimise the purchase of bought-in feedstuffs. There are many areas in the UK, for example, where the milk quota seems to have contributed to an increase in the intensity of grassland management, including the "improvement" and reseeding of old grassland areas which traditionally had been lightly grazed (NCC 1991). On the other hand, it is likely that the quota system has allowed some small dairy farmers to continue in production and thereby prevented the abandonment of some areas of grass.

Further incentives for intensifying production have come from a series of EC "structural" policies designed to increase agricultural productivity and encourage structural change in farming. These policies have included capital grants or similar support for new drainage works, irrigation projects, land improvement, the creation of larger holdings, the construction of farm roads, new land consolidation schemes, etc. Many of these policies have been aimed particularly at the more marginal agricultural areas and those classified as "less favoured areas" (LFA) under EC Directive 75/268 and its successors. The intention has been to encourage increased productivity and the emergence of more modern, generally larger holdings capable of providing a reasonable income in the longer term. In practice, much of this aid has not been able to reduce the disparities between richer and poorer farming areas and often has encouraged environmentally damaging intensification (Baldock 1989). Grassland reseeding is one of many forms of land improvement for which aid has been available, with EC contributions of 50 per cent or more in some countries.

The less favoured areas, as designated under the terms of Directive 75/268, now cover more than half the total agricultural land area of the EC. Within this huge zone, livestock farmers are eligible for special compensatory aids which usually take the form of "headage payments" per animal on the holding. There are very significant variations in the ways in which Member States make such payments. Some, such as the UK, pay close to the maximum sum permitted by the EC Directive; others, such as Spain, pay much smaller sums which are not available throughout the LFA. There are commensurate differences in the environmental impact of these policies. Comparing the use of headage payments in central Wales and the Auvergne in France in the early 1980s, Malcolm Smith found that there was a tendency to encourage over-stocking and over-grazing in mid Wales, but lower payments in France appeared to avoid this problem in the Auvergne, although not for reasons of nature conservation (Smith 1985). Generally speaking, headage payments are determined by reference to broad agricultural and budgetary goals and therefore are frequently insensitive to the optimum level of stocking required from an environmental point of view.

With the increasing concern about the accumulation of surpluses in the EC, there has been a growing tendency to offer farmers and other landowners incentives for afforestation, although there is still no formal EC forestry policy parallel to the CAP. There are now several forms of incentive for afforestation and a new proposal to encourage afforestation on set-aside land. In addition, some countries have received EC aid from outside the agricultural budget for forestry programmes. In Ireland, for example, a recent programme of afforestation, much of it targeted on peaty soils, has been financed partly by large loans from the European Investment Bank. Forestry programmes of this kind can encourage the conversion of marginal agricultural land to forestry, usually with non-indigenous species, such as sitka spruce in the UK and Ireland and pine and eucalypt species in Spain and Portugal. Afforestation is one of the main threats to marginal grasslands over extensive areas in Spain and Portugal.

Other policies which have encouraged agricultural intensification include the promotion of land consolidation schemes, the availability of low cost credit to farmers through state-aided institutions, state-financed research and development programmes, advisory services which are often partly or wholly state funded, and specific subsidies such as that on fertiliser use which was withdrawn in Spain only recently.

NEW EC POLICY MEASURES

Since 1985, new directions have been apparent in the CAP which has been concerned increasingly with the control of surplus production. During this period there has been greater acknowledgement of the environmental importance of Community agricultural policies and some steps have been taken, albeit modest ones, towards adopting more explicit environmental objectives for the CAP. While the preoccupation with surplus production and the need to control output has been crucial in opening the CAP to greater environmental influence, growing public concern, particularly in northern European countries, has also been an important factor. With the adoption of the Single European Act in 1987, which incorporates the new principle that "environmental protection requirements shall be a component of the Community's other policies", there is also a stronger legal foundation for greater integration between agriculture and environment policies.

Perhaps the most significant of these policies from the perspective of lowland dry grasslands is Article 19 of Regulation 797/85, as amended by 1760/87 and other measures. This allows Member States to provide assistance to environmentally sensitive forms of farming within defined areas and to recover part of the cost of doing so from the EC agricultural budget, FEOGA. Such payments are intended to compensate farmers for the loss of income which they experience as a result of adopting environmentally sensitive practices, which often include limits on fertiliser and pesticide use, restriction on stocking densities, restrictions on mowing and silage making, obligations to maintain landscape features and small habitats, and absolute prohibition on drainage, etc.

This approach has been adopted on a significant scale in the UK where 790,000 ha have been designated in 19 different environmentally sensitive areas, and in Germany where about 40 per cent of the entire land area of Bavaria, the largest Land (region) in Germany, has been designated. In the UK a large proportion of the environmentally sensitive areas (ESAs) are composed of grassland, although a relatively small percentage consists of chalk downland and other categories of lowland dry grassland. In Bavaria most of the land area affected is also grassland but the total includes extensive areas of wet meadows, alpine pastures, steeply sloping hills and relatively small areas of dry grassland. The only other countries which had utilised the Article 19 approach by early 1990 were Denmark and the Netherlands, although others were considering applications, including France, Ireland and Italy.

The relatively small-scale application of Article 19 in the EC can be explained by several different factors. One of these is the relative lack of interest in this approach by several governments, particularly in southern Europe where the emphasis is on increasing production and restraining land abandonment, rather than fostering more extensive systems of land use. In many countries, there is a lack of suitable institutions to draw up, promote and administer Article 19 schemes and often there is a lack of public pressure to support their introduction. Initially, the FEOGA contribution to Article 19 schemes was only 25 per cent and so the remaining three-quarters of the cost had to be found by national governments. This was a strong disincentive for the governments of the poorer Member States, although it has been partly overcome by recent amendments to the CAP structures policy which allow subsidies of up to 60-65 per cent in some of the poorer regions of the EC.

The Commission recently has proposed extensive amendments to Article 19 and part of another scheme to encourage agricultural "extensification"

(COM(90)366). Part of the argument for this amendment is that Article 19 should no longer be limited to designated sensitive areas, but should apply in the wider countryside and perhaps to broader categories, such as grassland or field boundaries. While this might allow a much wider application of the Article 19 approach, especially if more EC funds are earmarked for this policy, it is not entirely clear what the Commission intends. Under the current proposal, the aim seems to be to encourage a system of payments to farmers who reduce their inputs of fertilisers and agro-chemicals, perhaps by 20 per cent. All Member States would be obliged to introduce a measure of this kind, which would offer incentives for farmers undertaking to cut their input use, but it is less clear whether schemes to encourage the appropriate management of semi-natural vegetation would continue separately with EC aid or would be a category within the new reduced input scheme. Shortly before the end of 1990, the Commission suspended discussion on the draft of COM(90)366 because of the unresolved state of negotiations over the General Agreement on Tariffs and Trade (GATT). At the time of writing, it seems likely that new proposals to compensate farmers for further price cuts engendered by the GATT negotiations will include a new scheme similar to COM(90)366 but perhaps on a larger scale. It would be surprising if the current reform discussions do not result in at least one new scheme to pay farmers for more sensitive land management on a per ha basis.

If Article 19 has been implemented on only a limited scale, the EC's extensification policy has fared little better. A scheme was introduced under Article 1(b) of Regulation 797/85 whereby Member States are to introduce their own measures to offer farmers incentives for cutting production by 20 per cent or more without changing the area of land involved. This can be done either by requiring at least a 20 per cent fall in crop production or in the number of livestock on a holding or, alternatively, by a "qualitative method" which involves a change in the overall method of production, such as a switch from orthodox to organic farming. In either case, farmers are to be offered voluntary agreements over a five year period providing compensation for the income loss arising from reduced production.

This measure was agreed in 1987 and detailed rules were drawn up by the end of 1988. Initially Member States were required to do no more than introduce small scale pilot schemes. Even so, relatively few have done so. By early 1990, only Belgium, Germany, France and Italy had introduced extensification schemes. A pilot scheme for sheep and beef, on a very small scale, was announced in the UK in the summer of 1990. The only country to have adopted extensification on a significant scale is Germany. In the first year, 50,000 ha and almost 20,000 "livestock units" were enrolled in the German programme, with most of the farmers concerned electing to convert to organic systems of production. While this is a significant development in environmental terms, it is unlikely to have contributed very much to the conservation of lowland dry grasslands. During 1991 new proposals to amend this EC extensification scheme are expected from the Commission.

Extensification is seen as a complement to set-aside as a means of controlling surpluses. In principle, set-aside could be used to encourage the creation of new grassland habitats on land now in arable production. However, the current EC scheme (also under Regulation 797/85) is not designed in such a way as to provide farmers with much incentive for habitat creation. Set-aside contracts are for five years, whereas a longer period is necessary to re-establish most habitats of value for conservation. The scheme is not targeted so as to encourage the withdrawal of particular land from production although this would assist environmental goals, such as habitat creation. In most EC countries which have implemented set-aside much of the land is devoted to temporary or permanent fallow. Only in the UK and Germany have significant areas of land been withdrawn from production. However, the EC set-aside regulation is in the process of review and Member States may be able to introduce 20-year set-aside agreements if the regulation is amended. There are also plans to introduce EC incentives to encourage environmental management of set aside land, based partly on the Countryside Commission's own environmental top-up scheme in England.

These reforms of agricultural policy are being accompanied by new pressures to control water pollution from farming. One such pressure is the requirement on all EC Member States to comply with the 1980 drinking water Directive which sets standards for maximum nitrate concentrations in drinking water and also for maximum concentrations of pesticides. Nearly all Member States are having difficulty meeting the requirements of this Directive and in several northern European countries between two and five per cent of the population receive drinking water which is sometimes above the EC standard of 50 mg/litre. Consequently, many governments are introducing stronger controls on agricultural pollution and they are being encouraged to do so by further EC proposals to regulate the leaching of nitrate from farmland in areas where water supplies are most vulnerable to pollution. Most of these controls are concerned with the improved management of livestock wastes, including efforts to introduce maximum quantities of manure that can be spread on farmland, including grassland. None of these regulations is likely to result in sufficiently drastic reductions in nutrient inputs to greatly benefit species diversity on dry grasslands.

However, some of the new generation of farm pollution controls focus specifically on relatively small areas of agricultural land, usually land immediately surrounding groundwater boreholes or strips of land alongside water courses. Certain of these policies require a very substantial reduction in nutrient inputs to the land or, in some cases, the conversion of arable to extensively managed grassland. In the immediate vicinity of some boreholes, no inputs are permitted at all. Such policies in water protection zones, strips along streams and rivers, etc, will result in a growing area of land where inputs of nitrate and phosphate are reduced to a very low level or completely eliminated. There is clearly scope for encouraging appropriate forms of management on such land to increase species diversity and recreate different grassland and woodland habitats including lowland dry grasslands. This will require greater coordination between agencies responsible for water quality and those concerned primarily with nature conservation, which often function separately at present.

Such coordination could form part of a more environmentally sensitive approach to rural development. There is a growing emphasis within the EC on attempting to encourage rural development in a broader sense than simply increased agricultural output. This is evident in some of the recent proposals from the Commission's Directorate-General for Agriculture (DG VI), and in some aspects of the reformed EC Structural Funds which now include a more explicit commitment to rural development, partly targeted on areas designated as requiring particular assistance. At present, the new emphasis on rural development is still at an early stage and very often the environmental element is weak or non-existent, other than in countries where environmental policies are already well developed, such as in Germany. Nonetheless, in the longer term, rural development policies may provide a source of new Community money to assist environmental projects, including the maintenance of extensive agriculture in key areas. In some regions, such as Extremadura in Spain, proposals of this kind have been developed and it is to be hoped that they will attract support both from national governments and the Community.

Of more direct relevance to habitat management and creation is the proposed new EC Directive on wildlife species and their habitats which has been in draft form since 1988 but has yet to secure agreement by environment ministers. If adopted, this Directive would oblige EC Member States to designate and protect a considerable range of habitats, following rather similar principles to those found in the existing birds Directive. Resistance to this Directive from several Member States stems partly from a concern that they will be obliged to greatly extend their present network of protected areas and to adopt relatively expensive management policies. In a few countries where large areas of extensively managed semi-natural habitats remain, a considerable proportion of the land area could be involved; more than 30 per cent in the case of Spain for example.

Consequently, a need has been identified to create a new Community fund to assist countries in implementing the proposed habitats Directive and to strengthen their nature conservation arrangements. This has yet to be agreed but is likely to be an essential accompaniment to the habitats Directive.

POLICIES FOR GRASSLAND CONSERVATION AND RESTORATION

In reviewing the range of new policies recently adopted or emerging from the EC, it appears unlikely that many of the present agricultural proposals will play a central role in grassland conservation or restoration, other than the extension of the Article 19 approach over a larger area. Article 19 has been implemented in the UK, Germany, the Netherlands and Denmark by means of management agreements with individual farmers. Such agreements are capable of being sufficiently precise to maintain, or establish, the rather particular forms of management required to sustain semi-natural grassland habitats such as chalk downland, though they may overlook important factors such as grazing by rabbits. Many dry grassland habitats can be maintained only by suitable grazing patterns and stocking densities, which may need to be adjusted in the light of experience and monitored by a conservation agency with relevant expertise. While this precision can be achieved on individual farms, or even within small regions where the conditions are relatively uniform, it is difficult to attain by means of broad-brush incentives to extensify production, reduce input consumption by a specified amount or convert farm systems to an organic or other low input regime.

For this reason, management agreements are likely to be a central plank of any policy to protect and re-establish lowland dry grasslands, especially where these are semi-natural habitats dependent on specific grazing regimes. However, it is not essential that all such agreements stem from a single Community scheme or are financed entirely from a single budget. As already mentioned, there is scope for greater integration of water pollution control policies with habitat management objectives and this could be encouraged in the Community as well as at local and national level. Similarly, there is scope for developing habitat creation incentives within a reformed EC set-aside policy which should provide farmers with adequate incentives to withdraw production over 20 years or more, and permit some targeting of incentives to areas where withdrawal of land from production is particularly desirable from a conservation perspective.

In regions where significant areas of dry lowland grassland remain and support traditional systems of extensive agriculture, management agreements may not be the most appropriate form of incentive for farmers and land managers. For example, it may be possible to promote the products of local agriculture to consumers, emphasising both their quality and their ecological pedigree. In such a way it may be possible to attract a premium price and to provide incentives through the market for maintaining extensive agricultural systems. Rules to prevent intensification can be incorporated into this approach and enforced by local cooperatives, advisory services or other agencies. This approach is likely to be suitable only in a relatively limited number of areas in the short term at least, but may avoid some of the disadvantages of management agreements. Much less support from government agencies is required, for example, and the system may operate more easily in areas where much of the land is owned collectively or by local communities.

Incentives to maintain and create valued grassland habitats should be accompanied by measures to protect those that survive and prevent the encroachment of alternative more profitable land uses. The nature of the threats will vary from place to place, from region to region and from country to country. In some cases, the need is to strengthen land use planning and prevent the spread of urbanisation, waste disposal facilities, tourist developments and similar intrusions on to dry grassland. In others, there is a need to restrain agricultural improvement, commercial afforestation or other drastic changes in management. Several measures could play a role. One is the use of environmental impact assessment techniques and procedures, particularly for large scale afforestation. Another is the introduction of more sophisticated environmental screening processes within the Community, to ensure that aid from the EC Structural Funds is not devoted to environmentally destructive rural development and afforestation projects.

At the same time, it is necessary to maintain and strengthen nature conservation policies throughout Europe and to devote greater resources to habitat protection, especially in the poorer countries of southern and central Europe. The proposed EC habitats Directive and accompanying fund clearly could make a significant contribution.

In the longer term, it is possible to envisage more fundamental reforms in agricultural and rural policy in which the protection of rare habitats, such as dry lowland grassland, has a more central role. Several recent studies have proposed far-reaching reforms of the CAP for environmental purposes (Baldock 1990, Council for the Protection of Rural England 1990, Royal Society for the Protection of Birds 1990). For example, some have argued that in future farmers should receive prices for their products which are close to world market levels, and that subsidies should be offered in the form of land management payments on a per hectare basis. Environmental conditions would be attached to such payments. Others have emphasised the importance of linking all agricultural subsidies to some kind of environmental "cross-compliance". In effect, this would mean that farmers were eligible for public support, particularly price support, only if they complied with certain environmental obligations which would include habitat protection and regeneration. With the CAP currently under pressure from the US and other agricultural supporters through the GATT process, and also subject to further attempts to cut budgetary costs, radical environmental proposals may expect at least an airing even if the enormous pressure of vested interests makes it unlikely that the policy will be changed entirely.

In my own view, one important step forward would be to identify clearly those areas of lowland dry grassland and other important habitat types that remain, and prepare an authoritative European inventory which could then be expanded to include sites for habitat recreation. This would lay the foundations for more sensitive and clearly targeted plans.

REFERENCES

Baldock, D. (1990). *Agriculture and Habitat Loss in Europe*. WWF International, CAP Discussion Paper No. 3. Gland, Switzerland.

Baldock, D. (1989). *The CAP Structures Policy*. WWF International CAP Discussion Paper No. 2., Gland, Switzerland.

Bowers, J.K. and Cheshire, P. (1983). *Agriculture, the countryside and land use – An economic critique*. Methuen, London.

Commission of the European Communities (1990). *The Agricultural Situation in the Community*, 1989. Brussels.

Council for the Protection of Rural England/WWF (1990). *Future harvest*. London.

Frisch, J. (1988). National Report for Luxembourg. *In:* Park, J. R. (ed.): *Environmental management in agriculture – European perspectives*. Belhaven, London.

Green, B. (1990). Agricultural intensification and the loss of habitat species and amenity in British grasslands: a review of historical change and assessment of future prospects. *Grass and Forage Science*, 45: 365-372.

Nature Conservancy Council (1991). *Nature conservation and agricultural change*. Focus on Nature Conservation Paper No. 25. NCC, Peterborough.

OECD (1989). *Agricultural and environmental policies: Opportunities for integration*. OECD, Paris.

Smith, M. (1985). *Agriculture and nature conservation in conflict – the less favoured areas of France and the UK*. Arkleton Trust, Langholm, Scotland.

von Meyer, H. (1988). *The Common Agricultural Policy and the environment*. WWF International, CAP Discussion Paper No 1, Gland, Switzerland.

Royal Society for the Protection of Birds (1991). *Agriculture and the environment: Towards integration*, Sandy, England.

Wolkinger, F. and Plank, S. (1981). *Dry grasslands of Europe*. Nature and Environment Series No. 21. Council of Europe, Strasbourg.

MANAGEMENT OF SEMI-NATURAL LOWLAND DRY GRASSLANDS

John J Hopkins

English Nature, Northminster House, Peterborough PE1 1UA, UK

ABSTRACT

Semi-natural grassland is a plagioclimax habitat dependant upon management for its survival. A generalised analysis of the principles of semi-natural grassland management is developed, as appropriate to national and international perspectives, but the need for flexible policy frameworks for management emphasised, in order to take account of ecological variation within and between regions. It is suggested that more careful consideration of the relationship between vegetation structure and botanical and zoological features of conservation value is called for, in order to target management. Management systems compatible with nature conservation are seen as those in which nutrient capital is stable or declining and annual net primary production utilised. Grazing is seen as the most important management method, and the value of certain livestock management systems, as well as the desirable features of certain hardy breeds emphasised. Burning and cutting are seen as having much less importance and potential as management techniques in an agricultural context.

A number of recommendations for future policies are made, including the requirement to support economically and socially fragile farming systems which are most compatible with nature conservation management.

INTRODUCTION

In a European context the management of dry lowland grasslands is a complex issue. Such grasslands occur over a wide range of edaphic and climatic conditions, but most importantly the ways in which they are utilised by agriculturalists or managed by conservationists vary considerably, as do the social and economic systems affecting these land managers.

Much of what has been written about conservation management of grasslands has been directed towards the management of nature reserves, or similar situations where nature conservation is the main priority. A large part of this literature covers narrow subject areas, particularly where autecological problems are investigated, and such information is often only readily available, both in terms of location and content, to professional conservationists and ecologists.

It is the intention of the present author to attempt to develop a simplified analysis of grassland management, with the objective of providing a framework within which non-specialists involved with policies about nature conservation can be guided. In nearly all of this the assumption is made that the land manager will be a farmer. For current purposes the potentially valuable, even essential, role of nature reserves is not taken into account, and semi-natural grassland viewed principally in an agricultural context.

Optimal conservation benefit will arise from policies and strategies which accommodate regional and site specific variations in the ecological resource and its management requirements. However, it is not possible to produce this more detailed amplification of the management principles discussed here, where a British perspective of many management issues is reflected.

Additional information

Useful reviews of the theory and practice of grassland management for nature conservation are provided by Duffey *et al.* (1974), Ellenberg (1988), Green (1986), Hillier, Walton and Wells (1990), Rorison and Hunt (1980) and Wells (1990). A simplified guide to the farming of semi-natural grasslands is provided by ATB/NCC (1990), whilst Morris (1990) has reviewed the impacts of management upon invertebrates and Bacon (1990) provides valuable practical guidance about the management of grazing animals on nature reserves.

Nature conservation management of grasslands can be described as the low intensity agricultural utilisation of these habitats, even on nature reserves, and a substantial amount of relevant information occurs in the agricultural literature. However a majority of modern farming practices are directly damaging to grassland wildlife and such information needs treating with caution in its application.

PRINCIPLES OF GRASSLAND MANAGEMENT

Natural, semi-natural and artificial grasslands

Truly natural grasslands, seen here as ones which do not change to scrub or woodland in the absence of management, do occur in Europe, but only in the east were they ever extensive. Elsewhere they are confined to exposed coasts, mountains, skeletal or naturally toxic soils and other situations where harsh environments prevent tree and shrub growth.

Over most of lowland Europe grasslands are plagioclimax communities, which in the absence of management will gradually change to scrub and woodland. Such grasslands can be seen as falling into two types, semi-natural grasslands and what are termed here "artificial" grasslands. Semi-natural grasslands are composed of wild species which have not been deliberately introduced by man and have not been significantly modified by the use of agrochemicals, particularly artificial fertilisers. "Artificial" grasslands, which are not considered in detail below, have been sown by man, often using highly bred strains of grass and clover, or are derived from semi-natural grasslands by the application of artificial fertilisers and other deliberate or accidental means of nutrient enrichment.

In the lowlands of northwest Europe semi-natural grasslands account for only a very small part of the total grassland area, less than 3 per cent in the case of England and Wales (Fuller 1986). They do, however, exhibit a very wide range of floristic variation. In his classification of British vegetation, Rodwell (in prep.) identifies 16 communities and 49 sub-communities of grassland found on freely draining lowland soils in Britain. In part it is this wide range of ecological variation which, in Britain at least, makes simplified management formulae difficult or impossible to apply, for grassland communities vary considerably in their management requirements. Most notable in this context are those relatively rare grassland types managed over long periods as hay meadow, which contain species assemblages intolerant of summer grazing, a common practice in most other grasslands.

In contrast, a great majority of all artificial grassland belongs to one type, the *Lolio-Cynosuretum cristati*, found throughout large parts of Europe and other temperate zones of the world. Such grasslands contain a very limited range of now nearly ubiquitous temperate grassland plant species, which are highly competitive on nutrient rich soils, such as rye grass *Lolium perenne* and white clover *Trifolium repens*.

Vegetation structure

A familiar problem to many nature conservationists in Britain arises where features of botanical and zoological interest occur at the same site, as is often the case on dry lowland grasslands. When advice about management of such sites is sought from a range of specialists interested in different taxa or phyla, this is occasionally found to be incompatible.

Little attention has been paid to the resolution of this problem, and often the situation is explained as due to the differing sensitivities to management between taxa and phyla. This emphasis upon ecological differences does not, however, allow the conservation manager to resolve such conflicts of opinion, although the situation is here stated in an extreme form.

A subject about which consensus exists amongst zoologists is that structural features of vegetation are critical to the survival of many species and species assemblages, particularly where these occur at the edge of their range. What is seldom recognised is that amongst plant ecologists structural features of vegetation are also seen as critical determinants of vegetation floristics. The influential ideas of both Grubb (1977) and Grime (1973) place particular emphasis upon structural features of vegetation as influences upon higher plant reproduction and inter-specific competition amongst plants, key determinants of plant community composition.

At its simplest, grassland management is the manipulation of vegetation structure. It is by considering the way in which plants and animals respond to structural change induced by management that a more integrated approach to grassland management might be achieved. For example, in Britain not only is the stone curlew dependent upon a short grassland sward as breeding habitat, but this requirement is shared by the adonis blue butterfly *Lysandra bellargus* (Thomas 1990) and early spider orchid *Ophrys sphegodes* (Hutchings 1990). However other taxa of special conservation interest in Britain require long grassland swards for survival, such as the corncrake *Crex crex* and quail *Coturnix coturnix* amongst birds, Lulworth skipper butterfly *Thymelicus acteon* (Thomas 1990) and marbled white butterfly *Melanargia galathea* (BUTT 1986), as well as a wide range of hay meadow plants such as autumn crocus *Colchicum autumnale*, meadow crane's-bill *Geranium pratense* and globe flower *Trollius europaeus* (Hopkins 1990).

It is clear of course that there is no universally ideal structural type of vegetation. Rather, it is important that in each region account is taken of the range of structural types required to support the range of plants and animals occurring, and management regimes appropriate to this diversity sustained or promoted.

For animals, mixtures of structural types of vegetation may be important, as for example in the case of the woodlark *Lullula arborea* which in Britain requires scrub as nesting habitat, closely associated with short grassland where feeding takes place. However the boundary zones between different structural types of vegetation can also be important for rare and local plant species, notable in this context in Britain being a range of tall herb species such as bloody crane's-bill *Geranium sanguineum* and bastard balm *Melitis melissophylum* which occupy the scrub/grassland boundary, as do a number of invertebrates such as the Duke of Burgundy fritillary *Hamearis lucina* (BUTT 1986).

Plant nutrients and primary production

In an analysis of grassland management for nature conservation, Willems (1990) has seen historic precedent as a valuable guide, whilst Beintema (1988) has stressed the fundamental differences between modern intensive farming and nature conservation management of grasslands. Notably, both these authors work primarily in the Netherlands, where the semi-natural dry grassland resource is very diminished (Willems 1990). In other parts of Europe, including all regions of Britain, a major part of the surviving semi-natural grassland continues to be managed, not as nature reserves, but as part of farming systems, albeit often supported by government subsidies, which are increasingly directed towards nature conservation.

Not all these farming systems are capable of indefinitely sustaining the grassland of wildlife value. However, some can, and an important question to be asked is what common features are held by those historic and contemporary farming systems which have created and sustained semi-natural grasslands, and other semi-natural habitat types?

At the most fundamental level these systems have two important ecological properties in common:

1. The nutrient capital of the grassland ecosystem is low by comparison with artificial grasslands. The nutrient capital is stable, or in some cases gradually decreasing, due to the harvesting of plant and animal materials and largely influenced by natural inputs of nutrients due to aerial deposition and weathering of soil parent materials, as opposed to human activities.

2. Allowing for natural fluctuations influenced by climate, the standing crop of vegetation remains stable from year to year. The major part of annual net primary production is utilised either as animal feedstuff, or in other ways such as animal bedding.

Although the importance of sustaining stable structural features in grassland by manipulation of primary productivity is explored above, management of nutrients is of most immediate concern, for a major part of the damage and destruction of unimproved grassland which occurs is as a result of artificial nutrient enrichment, rather than direct physical destruction by ploughing (Fuller 1986). The effects of such nutrients are likely to be long-lived due to the low mobility of phosphorous in most soils. Indeed, several millenia after the last occupation, many human settlement sites can be identified due to residual phosphate enrichment, a technique of analysis commonly employed by archaeologists.

The way in which nutrient enrichment affects plant communities and thereby animals is complex. However, it is clear that in the majority of species-rich semi-natural grasslands, limited nutrient availability causes stress in plants, reducing growth rates, but favouring a rich assemblage of highly specialised "stress tolerant" species (Grime *et al.* 1990). The addition of nutrients reduces this stress, and results in the spread of a small number of highly competitive species capable of rapid growth, which replace the previously rich floristic assemblage and its associated dependent animals, as well as bringing about structural changes in the grassland (Hodgson 1986a, 1986b, 1987). Artificial fertiliser usage may not be the only way in which such nutrient enrichment takes place, however, as supplementary feeding of livestock can also result in gradual nutrient enrichment, particularly around feeding stations.

It should also be recognised that if management ceases or is reduced, and tall grass and scrub develop, this has important effects upon the nutrient ecology of a grassland site. Just as the phytomass increases in such successional change so does the nutrient capital, whilst the organic matter content of the soil will also increase. These changes have a long term effect at very many sites, for whilst it might be expected that cutting and removing scrub and re-establishing grazing or other management will re-create the conditions suitable for grassland wildlife to re-establish, the soils of the site may have undergone significant nutrient enrichment, resulting in the development of grassland communities typical of the nutrient enriched soils of modern agriculture (Green 1972).

The experimental work of Bakker (1990) illustrates the difficulty of removing increased nutrient loads in grassland soils. Nutrient enrichment may historically have been reversed in some cases by use of livestock to transfer nutrients from grasslands to arable land. Where sheep were held overnight in the now extinct management system of "folding", the animal dung deposited at night was valued as a fertiliser (Smith 1980). Moreover, soil erosion associated with cultivation seems to have played an important part in creating nutrient poor conditions in the soils of many semi-natural grasslands.

MANAGEMENT OPTIONS

There are three primary options open to the grassland manager: grazing, cutting and burning. These exhibit a number of fundamental differences in their ecological effects. In the case of grazing, defoliation of the sward is a gradual process which will most disadvantage those species with high palatability, whilst livestock trample the sward, and their dung and urine cause localised patterns of nutrient enrichment. By contrast, cutting acts suddenly and equally upon all plants, favouring those most able to respond to such treatment. Burning is also sudden and might be expected to favour those species with perennating structures held at or below the ground surface where they will be protected from scorching. It is significant that whereas grazing creates complex small-scale structural mosaics associated with variation in the palatability of individual species, as for example in rush *Juncus*-infested fields, these mosaics are destroyed by cutting and to a certain extent by heavy burning.

Further complication is introduced by the fact that not only can these management methods be applied in combination, but the intensity, frequency and seasonality with which they are carried out can be varied. This can provide a confusing array of options to the nature reserve manager.

However, in the context of farming practice the options would appear to be far more limited, due to the primary requirement to integrate these activities into a farming system. Further, in many instances problems now arise due to the abandonment of dry lowland grasslands, and it is important to understand the farming context in which grazing, cutting and burning are carried out if they are to be re-introduced onto these abandoned sites by the application of agricultural policies.

Grazing

By comparison with the artificial grasslands created by modern intensive agriculture, dry lowland semi-natural grasslands are for the most part low-productivity systems, which produce vegetation with relatively poor nutritional quality. Indeed it is to be expected that heavy grazing will act to increase the abundance of the least nutritious species due to selection of the most palatable species by livestock.

Despite this, grazing is the most frequent management practice carried out on dry lowland grasslands, in part due to practical difficulties of cutting such swards in an agricultural context.

Not all farming systems allow utilisation of semi-natural vegetation. Due to the agronomically undesirable features of the habitat, and faced with a re-structuring of European agriculture due to over-production and changing political systems, it is important to understand which types of farming are compatible with nature conservation management objectives, as these are not necessarily the most agriculturally productive or profitable. They are therefore particularly vulnerable to the effects of economic and political change.

As a very great generalisation, to take account of the entire European situation, it is possible to see pastoral farming systems as involving four principal types of livestock management. These are:

1. Dairying
2. Short-period, high-intensity fattening for meat
3. Long-period, low-intensity fattening for meat
4. Natural suckling

So far as dairying and the short-term fattening of animals for meat are concerned, these are practices which are generally incompatible with nature conservation management. In fact, many such animals are housed for long periods of their life and fed concentrated feedstuffs as part of their diet. This incompatibility arises because high production targets are set, which can only be met by high nutritional quality feed. Furthermore, breeds of animals concerned, which in Britain are nearly all cattle of the Frieisian, Holstein or Ayrshire dairy breeds, as well as crosses of these breeds with high productivity beef breeds, are unable to utilise low quality feedstuffs.

Where longer periods are allowed for fattening animals, utilisation of semi-natural grasslands may occur, although a period of enhanced feeding often takes place prior to slaughter to increase market value. Only a small amount of British beef is produced in this way at present, and where lowland dry grassland is utilised in such systems, it is usually in summer when the sward is most nutritious.

Of all systems of livestock management, the most important for nature conservation is that involving natural suckling, where female animals are maintained for a number of years, each year producing a crop of offspring which are then available for fattening. Both cattle and sheep may be involved and natural suckling means that low quality semi-natural grassland is converted by the mother into a high quality foodstuff, milk, for her offspring. After weaning, calves or lambs will often be moved from dry lowland grasslands onto artificial grasslands for fattening. However, at non-reproductive times of the year the female animal has only a minimal maintenance requirement of food and is still available for grazing of semi-natural grassland. Natural suckling systems therefore offer a long period of the year when livestock can be used for grazing of semi-natural grasslands within commercial farming systems.

As mentioned above, there are important differences in the animal breeds employed in these livestock production systems. Indeed, a fifth type of livestock production might be identified, that of maintaining breeds of livestock which have become rare due to low growth rate or other agriculturally undesirable characteristics. These breeds are attractive for grazing of semi-natural grasslands, as this system combines nature conservation with genetic resource conservation. Moreover, many of these breeds have strong regional associations and form part of a farming culture associated with semi-natural grassland exploitation. Certain rare breeds are particularly hardy and tolerate very low nutritional quality vegetation. They may also need minimal care at lambing or calving, and have resistance to common diseases. In the short term, however, the value of these hardy rare breeds is limited because they are not numerous and so can only be grazed on limited areas of grassland. Secondly, they may be difficult to integrate into economically viable patterns of farming even with subsidy, due to the strict market requirements of animal products.

Nevertheless, a substantial number of hardy animals suitable for grazing of semi-natural grassland persist in modern agriculture. In Britain, they are found primarily in the upland areas of the north and west, whilst elsewhere in Europe extensive areas of land which are intractable to agricultural improvement are likely to support similar hardy breeds with desirable qualities for nature conservation management. For example, in England, the Nature Conservancy Council has transferred the practices of grassland management in inhospitable upland areas to the lowlands, and made particular use on its chalk and limestone grassland reserves of Beulah sheep, a hardy Welsh hill breed. Other British breeds such as Herdwick, Swaledale, Scottish-blackface and Welsh mountain might equally have been employed. Significantly, such hardy breeds are readily marketable, if at lower profit, offering an opportunity to integrate nature conservation management into the market for animal products.

Cutting

In British agriculture, cutting of semi-natural grassland is now almost confined to two similar situations.

Firstly, cutting takes place to produce hay. However, this can only be carried out in situations where a tall sward develops, with high nutritional value. Such conditions occur only on the deeper more fertile soils, and it is in these situations where the most intensive farming practices are now carried out. Semi-natural grassland managed as hay meadow is now therefore rare in its occurrence.

The other situation where cutting took place in the past, but is now rare, is where unpalatable species such as bracken *Pteridium aquilinum* are abundant. In part this was carried out to provide a supply of material for use as animal bedding, thatch and other purposes. However, such cutting was also carried out to maintain good quality grazing land. Today, with the exception of hay meadows, cutting vegetation by farmers is a rare practice.

In Britain at least the opportunities for the expansion of cutting as a means of managing semi-natural vegetation seem limited. In all cases it is confined to level terrain over which machinery can be taken. The low standing crop of semi-natural grasslands means that it may be practically difficult to cut and gather the cut material. In addition, the feed value of the cut material would be unacceptably low for use in most stock rearing enterprises.

An alternative is simply to cut and leave material. This practice has no agricultural benefits and would have to be completely subsidised if it were to be carried out by farmers. It is also likely to be unacceptable ecologically, as the cut material will not only smother less competitive plant species and limit opportunities for regeneration from seed, but it is also likely at most sites to result in a gradual change towards the *Lolio-Cynosuretum* grasslands as repeated return of nutrients to the soil gives rise to seral eutrophication (Green 1972, 1986). There are indeed large areas of grassland cut in this way in British parks, gardens and urban green-spaces, almost none of which have features of special conservation value.

Burning

Burning grassland vegetation is uncommon in Britain, and was probably rare in the past, as it is directly destructive of potential animal foodstuff.

In all parts of the world where burning is carried out by farmers, it is done to improve the nutritional value of the sward by burning off dead or unpalatable material. In Britain such burning as does occur in dry grassland is mainly directed at the control of tor-grass *Brachypodium pinnatum*. On wetter soils, purple moor grass *Molinia caerulea* and rushes *Juncus* spp. are burnt occasionally.

In practice, unpalatable material is only a significant impediment to grazing British dry lowland grassland where neglect has allowed the build up of litter and in grassland invaded by tor-grass. Even in these circumstances, grazing with a suitable hardy breed of animal can give control of the situation without burning. However, it has to be recognised that such hardy animals are not found on all farms and burning therefore has a role to play in the restoration of grazing to neglected sites.

In theory, large areas of land which are currently ungrazed, and where there is no prospect of grazing or cutting, would be better managed by burning than by complete neglect. However the prospects for introducing extensive management by burning are limited in many highly settled parts of Europe due to the potential risk to property of uncontrolled burning. In Britain, more stringent controls on burning have been introduced partly due to the adverse effects of smoke and ash from stubble burning on human safety and quality of life. Successful control of fires requires a significant commitment of skilled labour and equipment.

CONCLUSION: AGRICULTURAL POLICIES AND GRASSLAND MANAGEMENT

A number of conclusions can be drawn about the types of agricultural policies necessary to conserve semi-natural grasslands, which might equally be applied to newly created grassland habitat:

1. Preventing further habitat destruction is the most immediate priority. However, unless this is associated with appropriate management in the medium- to long-term, habitat loss will not be prevented, simply delayed. *Policies need to consider semi-natural grasslands as a management system as well as an ecosystem.*

2. Pastoral farming is primarily concerned with the utilisation of primary production to create animal products, and to this end soil nutrient enrichment is carried out extensively outside of semi-natural grasslands. *Agricultural policies need to protect semi-natural grasslands against nutrient enrichment but promote the utilisation of their net annual primary production.*

3. There is no single management prescription suitable for all grasslands. *Future policies need to allow sufficient flexibility for regional and even site-specific management prescriptions to be implemented.*

4. Most grasslands are already managed as part of a farming system. *It is most important to protect and improve the features of these systems which are compatible with nature conservation objectives.*

5. The most valuable farming systems so far as nature conservation is concerned are those which involve the low intensity fattening of livestock and natural suckling of offspring. *These systems are economically and socially fragile and require special economic and other support targeted upon their environmentally desirable features.*

6. European farmers increasingly maintain breeds of livestock which have high productivity but are unsuitable for grazing semi-natural grasslands. *Incentives are required to protect and enhance the numbers of hardy breeds of livestock in the lowlands so as to allow management of semi-natural grasslands. The grazing of endangered breeds of livestock on semi-natural grasslands is culturally desirable.*

7. New areas of species-rich grassland will need to be managed in a manner similar to existing semi-natural grassland. *The chances of creating new grasslands successfully are limited until the problems of managing existing grasslands have been overcome.*

REFERENCES

ATB/NCC (1990). *Environmental handbook for agricultural trainers.* Agricultural Training Board, London.

Bacon, J.C. (1990). The use of livestock in calcareous grassland management. *In*: Hillier, S.H., Walton D.W.H. and Wells, D.A., *Calcareous grassland ecology and management*, pp. 121-127. Bluntisham Books, Huntingdon.

Bakker, J.P. (1989). *Nature management by grazing and cutting. On the ecological significance of grazing and cutting regimes applied to restore former species-rich grassland communities in the Netherlands.* Kluwer Academic Publisher, Dordrecht.

Beintema, A.J. (1988). Conservation of grassland bird communities in the Netherlands. *In*: Goriup, P. (Ed.), *Ecology and Conservation of Grassland Birds: ICBP Technical Bulletin No.7*, pp. 105-112. International Council for Bird Preservation, Cambridge.

Butterflies Under Threat Team (BUTT) (1986). *The management of chalk grassland for butterflies.* Nature Conservancy Council, Peterborough.

Duffey, E., Morris, M.G., Sheail, J., Ward, L.K., Wells, D.A. and Wells T.C.E. (1974). *Grassland ecology and wildlife management.* Chapman and Hall, London.

Ellenberg, H. (1988). *Vegetation of Central Europe* 4th edn. [English translation] Cambridge University Press, Cambridge.

Fuller, R.M. (1987). The changing extent and conservation interest of lowland grasslands in England and Wales: a review of grassland surveys 19301984. *Biological Conservation* 40: 281-300.

Green, B.H. (1972). The relevance of seral eutrophication and plant competition the management of successional communities. *Biological Conservation* 4: 378-384.

Green, B.H. (1986). *Countryside Conservation: the protection and management of amenity ecosystems.* 2nd Edn. George Allen, London.

Grime, J.P. (1973). Competitive exclusion in herbaceous vegetation. *Nature* 242: 344-347.

Grime, J.P., Hodgson, J.G. and Hunt, R. (1990). *Comparative plant ecology.* Hyman, London.

Grubb. P.J. (1977). The maintenance of species richness in plant communities: The importance of the regeneration niche. *Biological Reviews* 52: 107-145.

Hillier, S.H., Walton, D.W.H. and Wells D.A., (1990). *Calcareous grasslands ecology and management.* Bluntisham Books, Huntingdon.

Hodgson, J.G., (1986a). Commonness and rarity in plants with special reference to the Sheffield flora. 1: The identity, distribution and habitat characteristics of the common and rare species. *Biological Conservation* 36: 253-274.

Hodgson, J.G. (1986b). Commonness and rarity in plants with special reference to the Sheffield flora. 2: The relative importance of climate, soils and land use. *Biological Conservation* 36: 275-296.

Hodgson, J.G. (1987). Growing rare in Britain. *New Scientist*, 113 (No. 1547): 38-39.

Hopkins, J.J. (1990). British meadows and pastures. *British Wildlife* 1: 202-215.

Rodwell, J. (In prep.). *British Plant Communities.* Cambridge University Press, Cambridge.

Rorison, I.H. and Hunt, R. (1980). *Amenity grasslands an ecological perspective.* Wiley, Chichester.

Smith C.J. (1980). *Ecology of the English Chalk.* Academic Press, London.

Thomas, J.A. (1990). The conservation of Adonis blue and Lulworth skipper butterflies - two sides of the same coin. *In*: Hillier, S. H., Walton D.W.H. and Wells D.A., *Calcareous grasslands and management*, pp. 112-117. Bluntisham Books, Huntingdon.

Wells, T.C.E. (1989). Responsible management for botanical diversity. *In: Environmentally responsible grassland management.* 4.1-4.16. British Grassland Society, Hurley.

Willems, J.H. (1990). Calcareous grasslands in continental Europe. *In*: Hillier S.H., Walton, D.W.H. and Wells, D.A. *Calcareous grasslands ecology and management.* pp. 3-10, Bluntisham Books, Huntingdon.

RESTORING AND RE-CREATING SPECIES-RICH LOWLAND DRY GRASSLAND

Terry C E Wells

Institute of Terrestrial Ecology, Monks Wood Experimental Station, Abbots Ripton, Huntingdon, Cambridgeshire PE17 2LS, UK

ABSTRACT

Xerothermic grasslands in Europe have declined considerably over the past 40 years as a result of agricultural activities and are becoming increasingly fragmented and dispersed. With the current shift away from agricultural intensification, the opportunity exists to restore and recreate many semi-natural grasslands for the benefit of wildlife. Contrary to popular belief, not all species-rich grasslands are ancient: many have been formed by natural colonisation of previously arable land from adjacent seed sources. Whether this process can operate in the intensively farmed landscape of western Europe is open to question. Alternatives to natural colonisation, such as sowing mixtures of grasses and forbs, are considered and evaluated. The availability of suitable seed, the composition of mixtures and the use of hay-seeds and "nurse" crops are discussed. Soil fertility is recognised as one important factor in habitat creation, but the relationship between soil fertility and species diversity is still unclear and requires further research. Grassland management is essential for maintaining a desired structure and floristic composition, and again much more research is required into adapting agricultural methods of grassland management for nature conservation purposes.

INTRODUCTION

The xerothermic grasslands of Central Europe originated in prehistoric times as man cleared the primeval forest. All arid and semi-arid grasslands, as they are usually referred to in the European literature, are plagioclimax communities maintained by mowing, burning and grazing (Scherer 1925, Wells 1971, Ellenberg 1988). They are generally divided into two groups: the Festuco-Brometea, containing most of the calcareous grasslands and the Sedo-Scleranthetea composed of siliceous and generally acid grasslands.

Ellenberg's (1988) phytosociological classification of xerothermic grasslands (table 1), provides a broad framework into which they can be placed, the various sub-divisions reflecting differences in soils, temperature and moisture conditions. At the present time, the Festuco-Brometea occupy far larger areas of Central Europe than the siliceous grasslands, but all are under threat from agriculture and urban developments and are becoming increasingly fragmented and dispersed.

Agricultural activities are affecting semi-natural grasslands in a variety of ways: ploughing and cultivation destroys not only the vegetation but also inverts the soil profile leading to mineralisation of nitrogen and a temporary flush of nutrients; applications of large amounts of inorganic fertiliser will change chalk grassland (Mesobromion) from a species-rich community into a tall sward containing only a few highly productive grasses (Willems 1980). Even relatively small amounts of nitrogen, originating from atmospheric sources, have been shown to have an adverse effect on chalk grassland (Bobbink 1989), leading to floristic poverty and the dominance of *Brachypodium pinnatum*. As fragments of semi-natural grasslands become increasingly isolated in arable landscapes, management with grazing animals becomes difficult and successional processes occur which generally lead to dominance by a few grasses and invasion by shrubs. The speed at which these changes take place depends on the type of grassland, soil fertility and the nearness of invasive propagules: the end result is the loss of grassland.

With the current shift in the EC away from agricultural intensification towards favouring environmental protection and wildlife conservation, the opportunity exists to restore and recreate semi-natural grasslands. The aim of this paper is (1) to consider the ways and means of creating floristically-rich grasslands, (2) to identify the more important

Classes	Orders	Alliances
Festuca-Brometea (calcareous)	Brometalia (sub oceanic)	Xerobromion (arid)
		Mesobromion (slightly arid)
	Festucetalia valesiacae (sub-continental)	Festucion valesiacae (arid)
		Cirsio-Brachypodion (slightly arid)
		Koelerion glaucae (rich in lime)
Sedo-Scleranthetea (siliceous)	Corynephoretalia (sub oceanic)	Koelerion albescentis (on coastal dunes)
		Corynephorion (acid)
		Sedo-Scleranthetalia (sub-continental)

Table 1: Classification of xerothermic grasslands in Europe (after Ellenberg 1988)

factors to consider in selecting sites for habitat creation, (3) by means of selected case histories, show how recreation can be achieved and (4) to investigate management strategies for restoring grasslands. Attention will focus on chalk grasslands (Mesobromion), as this is the area where most research has been done, but the general principles which emerge should be relevant to other grassland types.

Re-creation by natural colonisation

Although Tansley (1939) described the chalk downlands of England as "having been used as sheep-walks for many centuries and probably from Neolithic times" and elsewhere, as "sheep-walks from time immemorial", there is now substantial evidence that many species-rich grasslands have originated from arable land abandoned at some time in the past (Smith 1980). In a detailed study of land-use on the Porton Ranges, Wiltshire, Wells *et al.* (1976) showed that differences in the time since land was abandoned resulted in different communities being formed which persisted for at least 130 years. While many chalk species were able to establish and persist in grasslands less than 100 years old, some species, notably *Asperula cynanchica*, *Carex caryophyllea*, *Filipendula vulgaris*, *Helianthemum chamaecistus*, *Helictotrichon pratense*, *Pimpinella saxifraga* and *Polygala vulgaris* were found only in grasslands more than 130 years old. Other species, such as *Carex humilis*, *Polygala calcarea* and *Thesium humifusum* were almost totally confined to old trackways where the shallow soils and dry conditions restricted the incursion of more aggressive species. In another study, Cornish (1954) demonstrated that species such as *Hypericum perforatum*, *Origanum vulgare* and *Plantago lanceolata* were able to invade ex-arable land within six to ten years. A similar suite of species was noted by Lloyd and Pigott (1967) as early colonisers of abandoned land in the Chilterns, although *Leontodon hispidus* and *Hieracium pilosella* were more prominent after about 20 years. Studies of chalk and limestone quarries (Davis 1976, Jefferson and Usher 1987), railway cuttings and embankments (Sargent 1984) and derelict industrial land (Greenwood and Gemmell 1978) indicate that under some circumstances species-rich communities may develop in a relatively short period of time.

More recent studies of the colonisation of ex-arable land have produced conflicting results. In a study into the use of sheep grazing to restore species-rich calcareous grassland, Gibson, Watt and Brown (1987) found that 43 of the 75 vascular plants restricted to patches of old calcicolous grassland within 2 km of the experimental site had colonised the field within two years. Most of these species had come from adjacent patches of old grassland but six came from further away. Species richness, diversity, and the abundance of individual plant species were increased by grazing in contrast to the control (ungrazed) areas. Grazing for a short time in spring produced more diversity and encouraged more species characteristic of permanent grasslands than did grazing for a short period in autumn, but the differences were slight. The results from this study suggest that provided seed sources occur near to the site, natural colonisation can be expected to produce a calcicolous grassland with many of the species characteristic of old chalk grassland, even though the structure and fine grain "texture" of the sward is different.

Quite different conclusions were drawn by Graham and Hutchings (1988a, 1988b) following an investigation of the size and composition of the seed bank beneath an improved chalk grassland adjacent to a species-rich grassland. Only 12 of the 49 species of seedling recorded from soil-cores taken over a two-year period from beneath the improved pasture were regular constituents of chalk grassland. Most of the species which germinated were normal components of the vegetation of a wide variety of habitats, and a high proportion were of major weed species present when the site was under cultivation. The seeds of chalk grassland species formed only a small part of the seed bank. A study of the seed rain falling on the site by Booth and Hutchings (1990) showed that seeds of chalk grassland species were dispersed only a few metres into the ex-arable field and that establishment was confined to the field edge. They concluded that for the majority of arable fields which are taken out of production in "set-aside" and similar Government supported schemes, the probability of re-establishing chalk grassland without re-seeding was low.

The results obtained from these two carefully conducted experiments emphasise the need for caution before making generalised statements concerning the establishment of semi-natural grasslands on ex-arable fields. Individual site factors, such as soil type, soil nutrient status, the nearness of appropriate seed sources, the direction of the prevailing wind and the management to which the newly created grassland is to be subjected (whether grazed, cut or unmanaged) are likely to be important and require further research.

Re-creation by sowing mixtures of grasses and forbs

Considerable experience has been gained in habitat creation since the mid-1970s when grass/forb mixtures were first sown in experimental plots in an attempt to create attractive grasslands (Wells, Bell and Frost 1981). Information on the performance of a range of species in seed mixtures has expanded as a result of monitoring long-term experimental plots, and through surveys of sites where wild flower mixtures have been sown (Wells 1985). We now know much more about the dynamics of changes in species composition of sown mixtures with time, especially under a constant management regime, but we are only just beginning to appreciate the importance of different management regimes on species composition.

Availability of seed

Prior to 1970, wildflower seed was available only in Europe, often collected by peasant farmers, and distributed through a complex network of small growers and agents. Some seed was imported from New Zealand and America, and it was often difficult to find out details of the seeds provenance. Since then the picture has changed dramatically. More than 90 per cent of the seed now sold in the United Kingdom is home produced, mostly by small specialist firms or by small growers producing seed for the larger seed companies. As the interest in habitat creation has grown, more attention has been given to using seed from local sources to avoid introducing "foreign" ecotypes into native populations. Volunteers from local conservation groups collect seed from local populations and grow these on to provide seed for local and regional restoration schemes. Seed of a wide range of grassland forbs is now available in quantity, most of it originating from native sources, and this should always be used in preference to imported seed.

Until recently, nearly all grass seed used in restoration work consisted of bred strains (cultivars) of native species. These were cheap, readily available in large quantities, but tended to be vigorous and in some cases highly competitive. Valuable information on the characteristics of new cultivars is given annually by the Sports Turf Research Institute, Bingley (Anon 1991). Some native grasses, such as *Trisetum flavescens*, *Alopecurus pratensis* and *Festuca ovina*, have proven to be less competitive and more suitable companions in wildflower mixtures and are slowly becoming available from commercial sources.

Composition of seed mixtures

The selection of species for inclusion in a seed mixture is partly a matter of personal choice and partly determined by what one is trying to achieve. Where the aim is to create a replica of an existing semi-natural grassland, such as chalk grassland, then the choice of species is largely determined by the composition of that grassland. This will show regional variation, not only on a European scale, but also within a country, and local botanists should be consulted as to the most appropriate "model" or "habitat stereotype" to use. Clearly, not all species can or should be included in a seed mixture - the seed of some will not be available, while it would make mixtures too expensive if all were included. Experience has shown that, however carefully a mixture is formulated, the species composition of the grassland will change within a few years: species not suited to the particular site conditions will disappear, while new (unsown) species will arrive. This process of "ecological sieving", which ultimately determines the composition of the grassland, is complex and it is unlikely that the mechanisms which control it will be understood in the short term. Some broad guidelines for mixtures, based on species which perform consistently well in the field, are given in Wells (1983). In general, rare or uncommon species should not be sown, as they mostly fail or do badly.

Experience has shown that a mixture composed of

Table 2: Grass/forb mixture suitable for creating a chalk or limestone grassland

Grasses (8)	% by weight of grasses	kg per ha	Cost (£)
Bromus erectus	10	2.4	24
Cynosurus cristatus	15	3.6	18
Festuca ovina	25	6.0	24
Festuca rubra ssp. *pruinosa*	20	4.8	24
Helictotrichon pratense	5	1.2	60
Koeleria cristata	5	1.2	60
Poa pratensis ssp. *angustifolia*	10	2.4	12
Trisetum flavescens	10	2.4	60
Total	100	24	282

Forbs (25)	% by weight of forbs	g per ha	Cost (£)
Achillea millefolium	4	240	18
Anthyllis vulneraria	4	240	38
Centaurea nigra	5	300	39
C. scabiosa	2	120	36
Clinopodium vulgare	2	120	55
Filipendula vulgaris	2	120	21
Galium verum	3	180	36
Leontodon hispidus	2	120	38
Leucanthemum vulgare	15	900	45
Lotus corniculatus	2	120	18
Medicago lupulina	2	120	1
Onobrychis viciifolia	5	300	25
Ononis spinosa	2	120	36
Origanum vulgare	2	120	30
Pimpinella saxifraga	1	60	12
Plantago lanceolata	5	300	20
P. media	5	300	21
Primula veris	4	240	48
Prunella vulgaris	10	600	30
Ranunculus bulbosus	2	120	10
Reseda lutea	4	240	22
Rhinanthus minor	10	600	60
Sanguisorba minor	4	240	31
Silene vulgaris	1	60	19
Stachys officinalis	2	120	18
Total	100	6,000	727

Mixture consists of 80% grasses to 20% forbs (by weight)
Grasses sown at 3,914 seeds per m², forbs at 894 seeds per m²
Sowing rate of grass forb mixture: 30 kg per ha (= 3.0 g per m²)
Sowing rate of Westerwolds Rye Grass: 10 kg per ha (= 1.0 g per m²)
Cost per ha (acre): £1,029 (416)

Table 3: Grass/forb mixture suitable for creating a meadow on calcareous clay soils

Grasses (9)	% by weight of grasses	kg per ha	Cost (£)
Agrostis capillaris	10	2.4	30
Anthoxanthum odoratum	2	0.5	38
Cynosurus cristatus	15	3.6	18
Festuca ovina	18	4.3	17
Festuca rubra ssp. *commutata*	15	3.6	12
Hordeum secalinum	5	1.2	84
Phleum nodosum	10	2.4	10
Poa pratensis	10	2.4	12
Trisetum flavescens	15	3.6	90
Total	100	24.0	311

Forbs (24)	% by weight of forbs	g per ha	Cost (£)
Achillea millefolium	4	240	18
Centaurea nigra	5	300	39
C. scabiosa	2	120	36
Clinopodium vulgare	2	120	55
Filipendula vulgaris	2	120	21
Galium verum	3	180	36
Hypericum perforatum	2	120	24
Leontodon hispidus	2	120	38
Leucanthemum vulgare	15	900	45
Lotus corniculatus	5	300	45
Lychnis flos-cuculi	1	60	22
Malva moschata	4	240	23
Pimpinella saxifraga	1	60	12
Plantago lanceolata	5	300	20
P. media	5	300	21
Primula veris	4	240	48
Prunella vulgaris	10	600	30
Ranunculus acris	10	600	36
Rhinanthus minor	10	600	60
Rumex acetosa	3	180	36
Sanguisorba minor	1	60	8
Saxifraga granulata	1	60	39
Silene vulgaris	1	60	19
Stachys officinalis	2	120	18
Total	100	6000	749

Mixture consists of 80% grasses to 20% forbs (by weight)
Grasses sown at 5,795 seeds per m², forbs at 1,133 seeds per m²
Sowing rate of grass forb mixture: 30 kg per ha (= 3.0 g per m²)
Sowing rate of Westerwolds Rye Grass: 10 kg per ha (= 1.0 g per m²)
Cost per ha (acre): £1,156 (468)

about 80 per cent grasses to 20 per cent forbs (by weight) will produce a grassland which has a similar structure and general appearance to long-established grasslands. There are sound reasons, both ecological and economic, why grasses should form the major component in any sward. In semi-natural grassland, grasses usually account for at least 60 per cent of the above-ground cover; they are able to withstand mowing, grazing and trampling and they remain winter-green. Many forbs, on the other hand, die down in winter and leave bare areas. Grasses form the matrix in which forbs and sedges are distributed.

Examples of mixtures suitable for use in lowland England for creating (a) a chalk or limestone grassland; (b) a calcareous clay meadow; and (c) an acid grassland are presented in Tables 2, 3, and 4.

These mixtures could be enhanced a year or more after they have been established by pot-grown plants of species which do not succeed when sown as seed, or for which only very limited supplies of seed are available. Examples of species which have been successfully introduced into chalk grassland using this method are *Briza media*, *Campanula glomerata*, *C. rotundifolia*, *Hippocrepis comosa* and *Thymus pulegioides*.

Use of hay-seeds to establish grasslands
This subject is dealt with fully in *Wildflower grasslands from crop-grown seed and hay-bales* (Wells, Frost and

Bell 1986). Floristically diverse grasslands can be created within two to four years using seed mixtures derived from species-rich alluvial meadows. Machines are now on the market (Plate 1) which can harvest seed direct from grasslands that are managed as hay-meadows, thereby creating an additional source of income for the farmer. Although most work so far has been on alluvial meadows, where results have been most encouraging, there is no reason why they should not be equally successful on chalk and limestone grasslands provided the topography is not too steep and obstructions, such as large ant-hills, are not present.

Use of a nurse crop

Many agricultural grasslands are "undersown" or sown with a nurse crop to ensure good establishment. The same technique has been applied with some success in establishing wildflower mixtures. Using a nurse crop has several advantages: (1) it germinates quickly and establishes a green vegetation (this may be of importance in providing cover for game birds), (2) it tends to suppress excessive annual weed growth and (3) it ameliorates harsh conditions and provides shelter for the slower germinating forbs.

Westerwolds rye-grass (an annual form of *Lolium multiflorum*), has been widely used as a nurse crop. Others, such as autumn sown cereals and some annual *Trifolium* and *Medicago* species may be equally suitable, but have yet to be tested with wildflower mixtures. Westerwolds is usually sown at about 3.0-4.5 g/m², the lower seed rate being recommended for diploid cultivars, the higher rate for tetraploid ones. Using Westerwolds on chalk and clay soils usually ensures an 80 per cent ground cover within 10 weeks of sowing. Provided it is cut before it sheds its seed, it will disappear from the sward within 2 years, enabling the sown perennial grasses and forbs to take its place.

Westerwolds is unlikely to perform well on infertile acid soils and, until more research is done on these soil types, mixtures should be sown without a nurse crop. On highly fertile soils, Westerwolds may pose problems because of excessive growth, but this may be overcome by cutting it two or three times in its first year for hay or silage. This strategy was used successfully on a fertile chalk soil at Great Chishill, Cambridgeshire, where a chalk grassland mixture was sown on land which had been used for continuous cereal production for the previous 20 years. Within one year of sowing, 23 of the 26 sown species had established.

Soil fertility and the re-creation of grassland communities

There has been much discussion recently concerning the relative importance of soil fertility in establishing species-rich communities on land coming out of arable production (Marrs and Gough 1989; Gough and Marrs 1990; Marrs, Gough, Griffiths and Grubb 1990). The need to consider the importance of high fertility has been highlighted by the new "set-aside" and "countryside premium" schemes designed to reduce agricultural surpluses, which provide unprecedented opportunities for habitat creation and vegetation restoration. It has generally been assumed that a high residual soil fertility brought about by the long-term use of fertilisers would impede the establishment of species-rich grassland communities on abandoned agricultural land.

Soil fertility and species diversity

The evidence that high levels of fertility reduces species diversity comes from three main sources, none of which are concerned directly with plant establishment on bare soil: (1) fertiliser experiments, such as those on the Park Grass plots at Rothamsted (Williams 1978, Silvertown 1980); (2) successional studies, where nutrient supply and soil fertility increase as the sere progresses, leading to reduced species-diversity; and (3) the effects of aerial pollutants, particularly the increased deposition of various sources of nitrogen, onto heathland (Diemont and Heil 1984) and chalk grasslands (Bobbink and Willems 1987, Bobbink 1989). These broad generalisations have often been accepted uncritically, especially by those concerned with set-aside, and have been used as an argument for not attempting

Table 4: Grass/forb mixture suitable for creating an acid grassland

Grasses (6)	% by weight of grasses	kg per ha	Cost (£)
Agrostis capillaris (or *A. castellana*)	10	2.4	30
Anthoxanthum odoratum	5	1.2	90
Cynosurus cristatus	20	4.8	24
Deschampsia flexuosa	15	3.6	108
Festuca ovina	30	7.2	28.8
Festuca rubra ssp. *pruinosa*	20	4.8	24.0
Total	100	24.0	304.8

Forbs (12)	% by weight of forbs	g per ha	Cost (£)
Achillea millefolium	4	240	18
Centaurea nigra	5	300	39
Galium verum	5	300	60
Leontodon hispidus	2	120	38
Leucanthemum vulgare	15	900	45
Lotus corniculatus	7	420	76
Malva moschata	10	600	57
Plantago lanceolata	10	600	39
Prunella vulgaris	10	600	30
Rhinanthus minor	15	900	90
Stachys officinalis	2	120	18
Trifolium dubium	15	900	4.5
Total	100	6000	514.5

Mixture consists of 80% grasses to 20% forbs (by weight)
Grasses sown at 6,292 seeds per m², forbs at 804 seeds per m²
Sowing rate of grass forb mixture: 30 kg per ha (= 3.0 g per m²)
Sowing rate of Westerwolds Rye Grass: 10 kg per ha (= 1.0 g per m²)
Cost per ha (acre): £839 (340)

Plate 1: The appearance of the plots two years after sowing on 26 May 1982; note red clover *Trifolium pratense*, ox-eye daisies *Leucanthemum vulgare* and meadow buttercups *Ranunculus acris*.

Plate 2: Eight years after sowing on 2 June 1986. Note numerous plants in flower (mostly meadow buttercup) and sweet vernal grass *Anthoxanthum odoratum*

Plates 1 and 2: Photographs of experimental plots of sown grassland

Table 5: Chalk grassland forbs which become established when sown as part of a seed mixture onto previously arable land

Achillea millefolium	*Lotus corniculatus*
Anthyllis vulneraria	*Medicago lupulina*
Campanula glomerata	*Onobrychis viciifolia*
Campanula rotundifolia	*Pimpinella saxifraga*
Centaurea nigra	*Plantago lanceolata*
Centaurea scabiosa	*Plantago media*
Clinopodium vulgare	*Primula veris*
Daucus carota	*Prunella vulgaris*
Filipendula vulgaris	*Rhinanthus minor*
Galium verum	*Scabiosa columbaria*
Helianthemum nummularium	*Taraxacum officinale*
Hieracium pilosella	*Thymus praecox*
Hippocrepis comosa	*Tragopogon pratensis*
Hypochoeris maculata	*Trifolium pratense*
Leontodon hispidus	*Trifolium repens*
Leucanthemum vulgare	*Veronica chamaedrys*
Linum catharticum	

the re-creation of floristically-rich communities. It is important that the following alternative point of view receives more attention:

(a) Not all successions lead to nutrient accumulation and a loss of species. In a detailed study of succession in temperate systems, Marrs *et al.* (1990) concluded "There is no clear relationship between soil fertility and successional stage. During some successions there is an increase in nutrient supplies, whereas on others there is either no effect or even a reduction in supply. As yet we do not know the reasons for these differences, but clearly there may be differential effects between species, sites and management histories".

(b) There is no convincing evidence for Britain that incoming atmospheric nitrogen is causing a decrease in floristic-richness in chalk grasslands. The effect of nitrogen inputs is confounded with management. Preliminary investigations into this relationship suggest that where high levels of grazing are maintained, floristic composition remains relatively unchanged.

(c) Floristically-rich grasslands are found on soils with a high total nutrient content: for example, some upland hay meadows, most of the remaining alluvial meadows in lowland river valleys which have not been fertilised and many ancient chalk grasslands. In these examples, most of the nutrients are "locked-up" in the organic matter and the amount of available nutrients is small.

(d) There is a strong positive relationship between soil organic matter and nutrient supply. Cultivated soils have a lower organic matter content than permanent grassland (Wells *et al.* 1976, Gough and Marrs 1990) and hence lower N and P contents. Nitrogen is known to be readily leached from unvegetated soils and therefore high residual levels of N in abandoned arable land are unlikely. Phosphorus is much less mobile in soils and becomes increasingly unavailable as pH increases. Where large amounts of phosphatic fertilisers have been used on crops in the past, phosphate levels in ex-arable soils may significantly exceed levels encountered in semi-natural grasslands (Gough and Marrs 1990). In some situations, this may lead to excessive growth by grasses and may give rise to problems in establishing forbs.

Reducing soil fertility

Concern about high nutrient levels in set-aside land has led to a consideration of the techniques available to reduce soil fertility (Marrs 1986, Marrs and Gough 1989). Except for sandy soils, nutrient depletion is likely to be a slow process unless drastic methods such as topsoil-stripping are employed. In practice, this solution to the problem is unlikely to be used, except in special circumstances, such as in urban areas where there might be a ready market for topsoil.

Experimental evidence suggests that species-rich chalk grassland can be established on "fertile" chalk soils. An experiment was sown in 1973 on an arable chalk soil in Hertfordshire which had previously grown 11 successive cereal crops. On the basis of a detailed soil analysis, the site could be classified as of intermediate fertility (extractable P, 1.43 mg/100 g, available N, 3.88 mg/100 g) in the scheme suggested by Allen *et al.* (1974). The experiment employed seven seed mixtures and a control (no seed sown) arranged in a randomised block design, replicated five times. During the first six months after sowing, the plots were cut twice to control arable weeds (of which 26 species were recorded, with *Sinapis arvensis* most abundant). In subsequent years, plots were cut in early August and again in mid-October, mowings being removed at the first cut but left on the ground at the second.

Of the 40 forbs sown in 1973, 33 had established by 1981 (Table 5) and are still there today, 17 years after the experiment began. Of the eight graminoids sown, six have established. In structure and floristic composition, many of the plots resemble a Mesobromion grassland in its fine structure and in the proportions of its constituent species it differs from some ancient Mesobromion grasslands, but these differences are apparent only to the connoisseur.

CONCLUSIONS AND RECOMMENDATIONS

Xerothermic grasslands of the class Festuca-Brometea have decreased throughout Europe and are becoming increasingly fragmented and widely dispersed. The consequences of this for all biota give cause for concern.

Historical studies suggest that not all species-rich grasslands are ancient and in previous times were probably formed by natural colonisation from nearby seed sources. In many countries, existing semi-natural grasslands are now uncommon and the only practical way of recreating some of these grasslands is to sow appropriate mixtures of grasses and forbs.

Wherever possible, native seed derived from local sources should be used in grassland restoration, to ensure the continued "integrity" of genetic resources. Where attempts are made to re-create replicas of existing grassland types, appropriate models should be chosen and results monitored at frequent intervals.

Soil fertility is an important factor to consider when selecting sites for re-creating grassland communities. Sites of low fertility, particularly those with low levels of available phosphate, are likely to be more suitable than sites with higher levels of nitrogen and phosphate, but the precise relationship between soil nutrient status and species diversity is unclear and more research is needed on this subject.

Management of both newly created and established grasslands is essential for maintaining their structure and floristic composition. Much more research is required on the effects of the range of techniques available for managing grasslands for nature conservation purposes.

REFERENCES

Anon. (1991). *Turfgrass Seed 1991*, Sports Turf Research Institute, Bingley.

Allen, S.E., Grimshaw, H.W., Parkinson, J.A. and Quarnby, C. (1974). *Chemical analysis of ecological materials*. Blackwell Scientific Publications, Oxford.

Bobbink, R. and Willems, J.H. (1987). Increasing dominance of *Brachypodium pinnatum* (L.) Beauv. in chalk grasslands: A threat to a species-rich ecosystem. *Biological Conservation* 40: 301-314.

Bobbink, R. (1989). *Brachypodium pinnatum* and the species diversity in chalk grassland. PhD Thesis, Utrecht.

Booth, K.D. and Hutchings, M.J. (1990). A study of the feasibility of re-establishment of chalk grassland vegetation following arable cultivation. *In:* Hillier, S.H., Walton, D.W.H. and D.A. Wells (Eds). *Calcareous grasslands - ecology and management, Proceedings of a BES/NCC Symposium at Sheffield University*, p.173, Bluntisham Books, Bluntisham.

Davis, B.N.K. (1976). Chalk and limestone quarries as wildlife habitats. *Minerals and the Environment* 1: 48-56.

Diemont, W.H. and Heil, G., (1984). Some long-term observations on cyclical and seral processes in Dutch heathlands. *Biological Conservation* 30: 283-290.

Ellenberg, H. (1988). *Vegetation ecology of central Europe*. Cambridge University Press, Cambridge.

Gibson, C.W.D., Watt, T.A. and Brown, V.K. (1987). The use of sheep grazing to recreate species-rich grassland from abandoned arable land. *Biological Conservation* 42: 165-183.

Gough, M.W. and Marrs, R.H. (1990). A comparison of soil fertility between semi-natural and agricultural plant communities: Implications for the creation of species rich grassland on abandoned agricultural land. *Biological Conservation* 51: 83-96.

Greenwood, E.F. and Gemmell, R.P. (1978). Derelict industrial land as a habitat for rare plants in S. Lancs. (v.c. 59) and W. Lancs. (v.c. 60), *Watsonia* 12: 33-40.

Graham, D.J. and Hutchings, M.J. (1988). Estimation of the seed bank of a chalk grassland ley established on former arable land. *Journal of Applied Ecology* 25: 241-252.

Graham, D.J. and Hutchings, M.J. (1988). A field investigation of germination from the seed bank of a chalk grassland ley on former arable land. Journal of Applied Ecology, 25: 253-263.

Jefferson, R.G. and Usher, M.B. (1989). Seed rain dynamics in disused chalk quarries in the Yorkshire Wolds, England, with special reference to nature conservation. *Biological Conservation*, 47: 12-136.

Marrs, R.H. and Gough, M.W. (1989). Soil fertility – a potential problem for habitat restoration. *In:* Buckley, G.P. (Ed.) *Biological habitat restoration*, pp. 29-44. Belhaven Press, London.

Marrs, R.H., Gough, M.W., Griffiths, M. and Grubb, P.J. (1990). *Reducing soil fertility on abandoned agricultural land for the restoration of native vegetation*. Report to NERC.

Sargent, C. (1984). *Britain's railway vegetation*. Institute of Terrestrial Ecology, Abbots Ripton, Huntingdon.

Scherrer, M. (1925). Vegetationsstudien im Limmattal Veroeffentlichungen des Geobotanisches Institut Ruebel 2. Zuerich.

Silvertown, J. (1980). The dynamics of a grassland ecosystem: botanical equilibrium in the Park Grass experiment. *Journal of Applied Ecology* 17, 491-504.

Smith, C.J. (1980). *The ecology of the English chalk*. Academic Press, London.

Tansley, A.G. (1939). *The British Isles and their vegetation*, vols 1 and 2. Cambridge University Press, London.

Wells, T.C.E. (1971). A comparison of the effects of sheep grazing and mechanical cutting on the structure and botanical composition of chalk grassland. *In:* Duffey, E. and Watt, A.S. (eds.). *The scientific management of animal and plant communities for conservation*. 497-515. Blackwell Scientific Publications, Oxford.

Wells, T.C.E. (1983). The creation of species-rich grasslands. *In:* Warren, A. and Goldsmith, F.B. (eds.). *Conservation in practice*. 215-232. Wiley, Chichester.

Wells, T.C.E. (1985). The establishment of floral grasslands. *Acta Horticulturae* 195: 59-69.

Wells, T.C.E., Sheail, J., Ball, D.F., and Ward, L.K. (1976). Ecological studies on the Porton Ranges: relationships between vegetation, soils and land-use history. *Journal of Ecology* 64: 589-626.

Wells, T.C.E., Bell, S.A. and Frost, A. (1981). *Creating native grasslands using native plant species*. Shrewsbury, Nature Conservancy Council.

Wells, T.C.E., Frost, A. and Bell, S.A. (1986). *Wild flower grasslands from crop-grown seed and hay-bales*. Peterborough, Nature Conservancy Council (Focus on nature conservation No. 15).

Willems, J.H. (1980). An experimental approach to the study of species diversity and above-ground biomass in chalk grassland. *Proc. K. Ned. Akad. Wet series C* 83, 270-306.

Williams, E.D. (1978). The botanical composition of the Park Grass Plots at Rothamsted 1856-1976. Rothamsted Experimental Station, Harpenden.

RELATED INITIATIVES FOR THE CONSERVATION OF LOWLAND HABITATS

Eric M Bignal

JNCC, Kindrochaid, Bruichladdich, Islay, Argyll PA44 7PP, UK

SUMMARY

This paper summarises items discussed at recent conservation initiatives which, although not primarily concerned with lowland dry grassland birds, are directly related through the broad land management issues that they address.

In 1988 an international workshop on the conservation of the chough *Pyrrhocorax pyrrhocorax* highlighted the overwhelming importance of traditional pastoral agriculture for choughs and other birds. A Wader Study Group workshop held in September 1989 considering the conservation requirements of waders breeding on wet grassland also identified the conservation value of low-intensity farming compared with high-intensity systems. As a result of the 1988 workshop, a European Forum on Birds and Pastoralism, which held its first meeting in November 1990, was established. This reviewed the importance and current status of low-intensity pastoral farming and discussed ways of preventing further decline.

All of these meetings have produced recommendations and the Nature Conservancy Council (NCC), Royal Society for the Protection of Birds (RSPB), World Wide Fund for Nature (WWF) and Institute for European Environmental Policy (IEEP) have individually and in combination produced statements on how future agricultural support mechanisms might favour long-established farming practices and their associated birds and other wildlife. An outline of some of these recommendations is given.

Recommendations for the conservation of birds of lowland dry grasslands should not be isolated from these other conservation initiatives. All are concerned with land management issues centred on maintaining long-established traditional farming methods. Since the pattern and character of farming is dependent upon political decisions at National, European and International levels there is potential for further collaboration to emphasise its importance for wildlife.

INTRODUCTION

This paper deals with "wider-countryside" areas and issues; that is the general countryside outside of sites where conservation management is the first concern. In particular it is concerned with the more remote and more natural parts of Europe, where the great majority of Europe's wildlife survives. In these areas individual sites can protect only a small proportion of the nature conservation interest (e.g. Stroud *et al.* 1990). Wildlife, and particularly the more mobile and wide-ranging species such as the larger mammals (wolf, pardel lynx, wildcat, wild boar, brown bear, deer etc.) and birds (eagles, kites, vultures, wading birds) depend for their survival on land management essentially concerned with agriculture. This is increasingly in association with forestry, tourism or a combination of these. Although for many years conservationists have been pointing to the habitat losses associated with agricultural intensification, conservation effort has tended to be occupied with ameliorating the damaging effects within intensively managed landscapes. Agriculture in this context has inevitably been seen as a negative influence. Little attention has been focused on the importance of certain forms of low-intensity extensive agriculture, based on stock rearing, in shaping and maintaining a characteristic and stable landscape in which wildlife has been able to survive and on which it now depends.

Perhaps two factors have served most to heighten awareness of the importance and fragility of these traditional farm systems.

The extension of the EC Common Agricultural Policy (CAP) into the more traditionally farmed areas of Europe in Spain, Portugal and Greece increased understanding of the need to support and protect farming systems over large tracts of countryside. Over-production of agricultural products has forced a review of the CAP and of support measures for European farmers. Reduction in commodity support because of over-production from intensively farmed areas and falls in the market prices of these commodities initially affects smaller traditional farmers hardest. This has resulted in an acceleration in the decline, which was already taking place, of low-intensity farming through either abandonment or intensification.

With the increasing awareness of inevitable changes in farm support structures from the EC and indications from politicians than more "environment friendly" farming would be used to maintain farmers incomes, NCC, RSPB, WWF, IEEP and others have made policy statements on the future of agriculture in Europe (NCC 1990, Baldock 1990, Taylor and Dixon 1990, Countryside Commission 1989).

Three international meetings held between 1988 and 1990 considered these issues in the context of bird conservation. In 1988 an international workshop on the conservation of the chough highlighted the overwhelming importance of traditional pastoral agriculture for choughs and other birds (Bignal and Curtis 1989). A Wader Study Group workshop held in September 1989 considering the conservation requirements of waders breeding on wet grassland also identified the value of low-intensity farming and the damaging effects of high-intensity systems to conservation (Hotker 1991). As a result of the 1988 workshop, a European Forum on Birds and Pastoralism was established holding its first meeting in November 1990. This reviewed the importance and current status of low-intensity pastoral farming and attempted to identify ways of preventing further decline (Bignal *et al.* 1991).

These meetings all discussed a number of points which may be of relevance to the final discussion session of this seminar. They can be summarised under three broad headings:

1. Ecological features and relationships — the basis for detailed management prescriptions
2. Farming systems and regional variations
3. Economics and politics — agricultural support

The degree to which the management related to (1), or other actions such as bird species action plans, can be implemented on the ground depends upon the survival of (2). The survival of (2) in nearly all cases will depend upon (3).

1. Ecological relationships

The ecological relationships between farm management, bird numbers, distribution and productivity are complex. Some key features and recommendations are outlined below. The full details can be found in the relevant publications:

(a) Common features of all areas are: open (often treeless) landscapes, a high proportion of natural pasture and high densities of grazing animals (often with seasonal movements), small patches of cultivation, arable stubbles, long pasture rotations, fallow land (often for up to 10 years).

(b) There is a complex patchwork of natural vegetation and cropped land in small compartments. "Untidy" features and micro-habitats which are lost with intensification survive, such as rocky outcrops, old field boundaries and earth banks, stone walls, uncultivated areas, ant-hills in pastures, permanently waterlogged soils.

(c) There are low, or negligible, inputs of non-organic fertilizers, herbicides and pesticides. Soil fertility is maintained using rotations and recycling animal dung. There is limited use of livestock veterinary chemical treatments which could have an insecticidal effect in dung (e.g. Ivermectin).

(d) Regional variations in farm management and natural vegetation are closely associated with bird distributions e.g. corncrakes with marshes and traditionally harvested hay crops in the Hebrides, wading birds with saltmarsh and lowland wet meadow grasslands, choughs with cliffs and extensive stock rearing in western Britain and Ireland and the mountain areas of Iberia.

(e) Permanent pasture grasslands in low-intensity systems have richer invertebrate faunas than short-lived, heavily fertilised pastures. The complex vegetation structure resulting from mixed farming provides feeding and nesting conditions not found on intensively managed farmland.

(f) The meetings identified the functional as well as the physical importance of these features. The "functional unit concept" is useful in understanding the full requirements of bird populations (Tamisier 1979, 1985, Bignal *et al.* 1989, Stroud *et al.* 1990).

(g) Results of long-term monitoring studies are essential to understanding how breeding density relates to breeding production. There is evidence for some species that breeding birds may be "trapped" in the early part of the season into nesting in areas where subsequent chick and nestling survival is low (e.g. corncrake, choughs and waders).

2. Regional variations in farming systems

Although there are common features across the low-intensity pastoral farmland in Europe there are nevertheless strong regional variations in farm systems reflecting national policy as well as the direct effects of climate, geology and topography. Pastoral-based low-intensity agriculture in the EC can be categorised as transhumance or sedentary. Nomadic pastoralism of the type found in North Africa and Asia does not occur.

Two examples of regional systems discussed at the 1990 Forum were transhumance in Iberia and crofting in the Scottish Highlands.

Transhumance once formed the basis of stock rearing in much of Europe from the Scottish Highlands (Bil 1989) to the Mediterranean lands.

Historically it was developed most extensively across the Iberian peninsula (Ruiz and Ruiz 1989). Although much reduced it still survives as a viable agricultural system in Spain and its survival, and re-establishment, will be crucial for many birds such as chough, bustards, vultures and eagles. This long-established agricultural system has been fundamental in shaping the mountain pasture grasslands of the Central mountains and the Pyrenees, the *dehesa* wood-pastures of the Meseta and the *cañadas* (drove roads) that link them (Ruiz and Ruiz 1989).

Crofting in the Scottish Highlands and Islands is a good example of a low-intensity sedentary livestock rearing system. It has many similarities with small scale farming systems of southern Europe. Characteristic elements of these systems are relatively large areas of grazing land (often "common pastures") and patchworks of crop production for animal fodder and human consumption (Fuller *et al.* 1986, Bignal *et al.* 1988, Stroud and Pienkowski 1989). There is a complex organisational structure which maintains the systems and human communities in rural areas.

Conservation initiatives must address the question of maintaining farm systems if detailed management prescriptions are to have long-term benefits.

3. Economics, politics and agricultural support

The Common Agricultural Policy (CAP) is of overwhelming importance in directing national agricultural policies, it is the basic framework of support for agriculture in the European Community. The CAP itself is to a degree shaped by even wider, non-EC, pressures through continuing negotiation on the General Agreement on Tariffs and Trade (GATT).

Since 1988 pressures to reduce the cost of the CAP, to reduce over-production and to remove price barriers around the EC, have led to increasing efforts to reduce price support and replace this with support to strengthen the social and environmental benefits of agriculture (NCC 1990).

In the light of this background a number of recommendations were made by the Forum for consideration in CAP reviews and elsewhere:

(a) EC Member States should urgently review the suite of Special Protection Areas (Directive 79/409) and other mechanisms in the wider environment.

(b) The EC Habitats Directive needs to be complemented by action under the CAP to widen its application to traditional forms of agricultural land management.

(c) EC Member States should review the implementation of the Less Favoured Areas Directive (75/268) so that it assists, as originally intended, farming which is necessary to protect the countryside. In particular the Environmentally Sensitive Areas concept (ESA) should be used to protect farming systems. New ESA measures should grade designated areas to provide more support in areas of highest interest. In the best areas payments for maintaining the status quo should be developed.

(d) Within the UK, Less Favoured Areas (LFA) should be reviewed and redefined to delimit those areas where traditional farming survives. Much of the current LFA have been intensified and no longer retain their wildlife, social or landscape characteristics.

(e) In order to be effective at the wider, farm-system scale it is necessary to identify areas of particular conservation importance. For example, these meetings targeted the Scottish Hebrides (crofting and stock rearing), the French and Spanish Pyrenees, the pre-Pyrenean hinterland and the Bardenas Reales (transhumance) and the Spanish Asturian mountains, northeast Portugal (Tras os Montes) and the Portugese Serra da Estrella.

(f) The CAP support system should provide price support for crop or livestock production methods with environmental or social benefit but which are currently discriminated against. A clear example of this is hay production on small farms. Throughout the EC economic pressures are resulting in either intensification (into silage production) or abandonment.

(g) In the UK measures should be developed to encourage the re-establishment of the traditional stratification of livestock production (between lowland, upland and hill farms), for instance by limiting the number of breeding animals eligible for subsidy in the lowlands.

CONCLUSIONS

Despite the enthusiastic, and increasingly widespread, support within the conservation lobby for agricultural measures which embrace these principles, reaction in the UK from the Agriculture Minister and NFU (England and Wales) to recent proposals to reform the CAP to favour smaller farmers (MacSharry, February 1991) has been disappointing. The overall prognosis for these initiatives in the immediate future is pessimistic. The likelihood of a real shift in emphasis of agricultural support away from production (quantity) towards a more broad-based, beneficial type of land management (quality) seems slim. They point to a continuing lack of acceptance amongst agricultural policy makers of the environmental and social benefits of traditional agriculture. Perhaps this is inevitable since primary policies aimed at low intensity farming, although having great environmental and social benefit, will not benefit the larger, more intensive farmer. There continues to be a fundamental difference in perception between the UK farmers union and the agricultural departments on the one hand and conservation managers on the other. The former continues to see its main contribution to the environment in redressing the damage caused by intensive farming rather than specifically maintaining farming systems which are currently beneficial. This is associated with the view that all farmland is of some wildlife interest and is perhaps inevitable at a time when all possible methods of maintaining farmers incomes are being explored. We should make it clear that not everything that farmers (even low-intensity farmers) do is good for wildlife. Rather, there is an opportunity to shift the emphasis intentionally to provide favourable management, rather than, as currently, by default.

There is a lack of acceptance of the need to diversify within intensive farmland and not out of farming in extensive low-intensity farmland.

There is a real danger that when set against the search for introducing financial payments for incorporating "green" or "environmental" management into intensively managed farmland, the needs of the smaller, less mechanised, more remote, more traditional farmers will be overlooked - or at the best tacked-on as an afterthought.

A main conclusion of the meetings is that conservation bodies need to emphasise the environmental benefits of both supporting existing low-intensity extensive farming systems and of extending these systems into areas currently farmed intensively. This would have social, landscape and wildlife benefits. It would mean however that some large intensive farms could no longer survive. It would shift the pressure from the smaller, often part-time farmer, to the large agri-business. Whether this is a realistic situation remains to be seen.

ACKNOWLEDGMENTS

I would like to thank Professor David J. Curtis and David Stroud for commenting on the draft of this paper. Much of the content derives from the discussion sessions at the Second European Forum on Birds and Pastoralism held at Port Erin, Isle of Man, October 1990.

REFERENCES

Baldock, D. (1990). *Agriculture and habitat loss in Europe*. WWF International CAP Discussion Paper Number 3.

Bignal, E.M., Curtis, D.J. and Matthews, J.L. (1988). *Islay: Land types, bird habitats and nature conservation. Part 1: Land use and birds on Islay*. NCC Chief Scientist Directorate Report No. 809, Part 1.

Bignal, E.M. and Curtis, D.J. (eds.) (1989). *Choughs and land-use in Europe*. Scottish Chough Study Group, Tarbert.

Bignal, E.M., Bignal, S. and Curtis, D.J. (1989). Functional unit systems and support ground for choughs — the nature conservation requirements. *In:* Bignal, E.M. and Curtis, D.J. (eds.). *Choughs and land-use in Europe*. Scottish Chough Study Group, Tarbert.

Bignal, E.M., Curtis, D.J. and Curtis, M.A. (1991). *Birds and pastoral agriculture in Europe*. Scottish Chough Study Group.

Bil, A. (1989). Transhumance economy, setting and settlement in Highland Perthshire. *Scot. Geog. Mag.* **106**(3): 158-167.

Countryside Commission (1989). *Incentives for a new direction for farming*. CCP 262, Cheltenham.

Curtis, D.J. and Bignal, E. (1991). *The conservation role of pastoral agriculture in Europe. A Discussion Document*. Scottish Chough Study Group, Tarbert.

Fuller, R.J., Reed, T.M., Buxton, N.E., Webb, A. Williams, T.D. and Pienkowski, M.W. (1986). Populations of breeding waders *Charadrii* and their habitats on the crofting lands of the Outer Hebrides, Scotland. *Biological Conservation* **37**: 333-361.

Hotker, H. (ed.) (1991). Waders breeding on wet grassland. Proceedings of a workshop held at the Wader Study Group Conference in Ribe, Denmark on 23 Sept. 1989. *Wader Study Group Bull. 61, supplement*.

Nature Conservancy Council (1990). Nature Conservation and Agricultural Change. *Focus on Nature Conservation No. 25*. Nature Conservancy Council, Peterborough.

Ruiz, M. and Ruiz, J.P. (1986). Ecological history of transhumance in Spain. *Biological Conservation* **37**: 73-86.

Stroud, D.A. and Pienkowski, M.W. (1989). The conservation importance of Coll and Tiree: current land-use and future prospects. Pp. 143-148. *In:* Stroud, D.A. (ed.). *Birds on Coll and Tiree: status, habitats and distribution*. NCC/SOC, Edinburgh.

Stroud, D.A., Mudge, G.P. and Pienkowski, M.W. (1990). *Protecting internationally important bird sites: a review of the EEC Special Protection Area network in Great Britain*. Nature Conservancy Council, Peterborough.

Tamisier, A. (1979). The functional units of wintering ducks: a spatial integration of their feeding and habitat requirements. *Verh. orn. Ges. Bayern* **23**: 229-238.

Tamisier, A. (1985). Some considerations on the social requirements of ducks in winter. *Wildfowl* **36**: 104-108.

Taylor, J.P. and Dixon, J.B. (1990). *Agriculture and the environment: towards integration*. RSPB, Sandy.